LABORATORY MANUAL
to Accompany

Electronics Technology Fundamentals

Third Edition

Robert T. Paynter

B.J. Toby Boydell

PEARSON
Prentice
Hall

Upper Saddle River, New Jersey
Columbus, Ohio

Editor in Chief: Vernon Anthony
Acquisitions Editor: Wyatt Morris
Editorial Assistant: Christopher Reed
Production Coordination: Shelley Creager, Aptara®, Inc.
Project Manager: Rex Davidson
Senior Operations Supervisor: Pat Tonneman
Operations Specialist: Laura Weaver
Art Director: Candace Rowley
Cover Designer: Diane Lorenzo
Cover Photo: iStock
Media Editor: Michelle Churma
Media Project Manager: Karen Bretz
Director of Marketing: David Gesell
Marketing Assistant: Les Roberts

This book was set in Times Roman by Aptara®, Inc. and was printed and bound by Edwards Brothers Malloy.

Pearson Education Ltd., London
Pearson Education Singapore Pte. Ltd.
Pearson Education Canada, Inc.
Pearson Education—Japan

Pearson Education Australia Pty. Limited
Pearson Education North Asia Ltd., Hong Kong
Pearson Educación de Mexico, S.A. de C.V.
Pearson Education Malaysia Pte. Ltd.

10
ISBN-13: 978-0-13-504876-4
ISBN-10: 0-13-504876-1

Preface

This laboratory manual has been written to accompany the third edition of *Electronics Technology Fundamentals* by Robert T. Paynter and B.J. Toby Boydell. As such, the exercises have been arranged (as closely as possible) to follow the progression of topic coverage in the text. Exercises 1 through 12 deal with dc circuits and principles, and Exercises 13 through 27 deal with ac circuits and principles. Exercises 28 through 48 deal with discrete devices and circuits, op-amps, and op-amp circuits.

SIMULATION EXERCISES

The simulation circuits for this lab manual were created for use with Multisim® software and are provided at no extra cost to the consumer. Multisim is a schematic capture, simulation, and programmable logic tool used by college and university students in their course of study of electronics and electrical engineering. Multisim is widely regarded as an excellent tool for classroom and laboratory learning. To access the files, go to www.prenhall.com/paynter.

The exercises in this manual have been tested thoroughly. At the same time, we recognize that there is no such thing as a "universal" lab. That is, we realize that not every exercise will work the same way for every student in every school. Therefore, we have included the results of circuit simulations for each exercise in the *Online Instructor's Resource Manual*. These simulation results, while not absolute, will give you a good idea regarding the type of results your students should obtain. We hope that the simulation results in the resource manual, along with the lab manual itself, make for a great learning experience in the lab. To access supplementary materials online, instructors need to request an instructor access code. Go to **www.pearsonhighered.com/irc**, where you can register for an instructor access code. Within 48 hours after registering, you will receive a confirming e-mail, including an instructor access code. Once you have received your code, go to the site and log on for full instructions on downloading the materials you wish to use.

Parts List

(*Note:* Quantities refer to the maximum number required for any single exercise.)

RESISTORS (1/4 W)

Quantity	Value	Quantity	Value	Quantity	Value
1	10 Ω	4	1.0 kΩ	3	10 kΩ
1	22 Ω	1	1.1 kΩ	1	12 kΩ
1	47 Ω	1	1.2 kΩ	2	15 kΩ
2	100 Ω	1	1.5 kΩ	1	16 kΩ
1	220 Ω	1	2.0 kΩ	1	22 kΩ
1	330 Ω	2	2.2 kΩ	1	27 kΩ
1	390 Ω	1	2.7 kΩ	1	30 kΩ
1	470 Ω	1	3.3 kΩ	1	33 kΩ
1	510 Ω	1	4.7 kΩ	1	39 kΩ
1	560 Ω	1	5.6 kΩ	1	47 kΩ
		1	6.8 kΩ	1	68 kΩ
		1	8.2 kΩ	1	82 kΩ
				1	100 kΩ
				1	330 kΩ
				1	1 MΩ

RESISTORS (POWER)

Quantity	Value
1	500 Ω (5 W)
1	10 kΩ (2 W)

POTENTIOMETERS

Quantity	Value	Quantity	Value	Quantity	Value
1	1 kΩ	1	10 kΩ (linear taper)	1	2 MΩ
1	2 kΩ	1	10 kΩ (audio taper)	1	5 MΩ
1	5 kΩ	1	10 kΩ (10-turn)		
		1	25 kΩ		
		1	50 kΩ		
		1	100 kΩ		

CAPACITORS AND INDUCTORS

Quantity	Value	Quantity	Value	Quantity	Value
1	51 pF	1	0.1 μF	1	1 mH
1	100 pF	1	1 μF	1	10 mH
1	1 nF	2	10 μF		
1	2.2 nF	1	22 μF		
2	10 nF	2	47 μF		
1	22 nF	1	100 μF		
		1	470 μF (50 V)		

DEVICES

Quantity	Part Number	Description
4	1N4001	Rectifier diode
2	1N4148	Small-signal diode
1	1N5240	Zener diode
1	T1	Red LED
2	2N3904	*npn* transistor
1	2N3906	*pnp* transistor
2	2N5485	*n*-channel JFET
1	2N5060	Silicon-controlled rectifier (SCR)
2	KA741	Operational amplifier
1	LM555	Timer IC
1	LM317	Voltage regulator (TO–220 case)

MISCELLANEOUS

Quantity	Description
1	Decade resistor box
1	Transformer, 25.2 V center-tapped
1	Momentary push-button switch (NO)
1	Soldering iron
1	Heat sink for TO–220 case
1	Prototyping board
1	¼ A fuse
1	½ A fuse
1	Fuse holder

TEST EQUIPMENT

Quantity	Description
2	Digital multimeters (with appropriate leads)
1	Volt-ohm-milliammeter (with appropriate leads)
1	Dual-trace oscilloscope
2	Oscilloscope probes (should have ×10 capability)
1	Variable dc power supply (dual-output)
1	Function generator (with dc-offset capability)

SPECIFICATION SHEETS

1N4001	Rectifier diode
1N5240	Zener diode
2N3904	*npn* transistor
2N5485	*n*-channel JFET
KA741	Operational amplifier
2N5060	Silicon-controlled rectifier

Contents

To the Student

SAFETY

Like most technologies, electronics can be hazardous if approached carelessly or without proper training. The circuits in this manual are designed for relatively low currents, but even low-current circuits can harm you under certain circumstances. To keep the laboratory environment as safe as possible, always follow these guidelines:

1. Always ask your instructor for help if you are unsure of how to proceed or use a specific piece of equipment.
2. Always read the exercise *before* coming to lab. By doing so, you can identify any concerns before beginning.
3. Until directed otherwise, have your instructor inspect your circuits before applying power.
4. Use only equipment that is approved by your instructor.
5. Always disconnect circuits before leaving your work station. (This holds true for soldering irons as well.)
6. Unless directed otherwise, perform all the steps in each exercise in their proper order.
7. *Never try to defeat any safety feature on your equipment or in your circuit.*
8. Never touch the components in a live circuit unless specifically told to do so. Even when power is disconnected, some components, such as capacitors, can store a significant charge.
9. Never bring food or liquids into your work area.
10. Avoid wearing jewelry, such as rings and watches, when working on live circuits.
11. Never work alone.
12. Don't make the mistake of believing that safety rules are written for everyone *else* to follow. They are written for *you*.

GETTING THE MOST FROM THE EXERCISES

Here are some suggestions that will help you to have a positive and educational experience in the laboratory:

1. Come prepared. Lab time is limited, so the more prepared you are, the less lab time you will spend trying to get ready.

2. Review the text material identified under the *LAB PREPARATION* heading before coming to lab.

3. Perform any relevant calculations before coming to lab. That way, you'll have an idea of the results you should expect before performing the exercise.

4. Make certain that you have all the needed components and equipment before starting the exercise.

5. Work in a neat and organized manner. Whenever possible, build your circuits so that they are laid out exactly like the schematic. That way, you'll never have trouble identifying the component of interest for a specific step.

6. Record your results when you take a measurement. Don't rely on memory to "fill in the blanks" later.

7. Use your own words to answer the questions in each exercise, making your answers as complete as possible. Here's a simple guideline: Answer the questions as if you were trying to explain the operating principle, measurement, or calculation to someone.

If you follow these suggestions and the safety rules presented earlier, your experience in the lab will be beneficial and educational rather than a source of frustration.

Part 1

DC Principles

Exercise 1

The Resistor Color Code
and Resistance Measurements

OBJECTIVES

After completing this exercise, you should be able to:

- Determine the nominal value of a resistor using its color code.
- Measure resistance using a *digital multimeter* (DMM) and an analog *volt-ohm-milliammeter* (VOM).
- Determine the tolerance of a resistor using its color code.
- Calculate the percent of variation between a nominal value and a measured value.

DISCUSSION

Resistors are components that provide opposition to current. They are one of the most common components in any electronic circuit and come in a variety of shapes, sizes, and materials. Some have fixed values, while others have values that can be varied within predetermined limits.

Resistors come in a variety of rated (or *nominal*) values. One of the most common ways of indicating resistor values is the *resistor color code*. In this exercise, you will use the resistor color code to determine the rated values for a variety of resistors. You will then compare the rated component values to their measured values.

Resistors, like all man-made devices, are not perfect. The actual (measured) value of a resistor may vary from its nominal value. The *tolerance* of a resistor indicates the maximum range over which the actual component may vary, given as a percentage of the nominal value. For example, a 1 kΩ resistor with a 2% tolerance has a value of 1 kΩ ±2% under normal circumstances. In this exercise, you will learn how to calculate the *percent of variation* between nominal and measured resistor values.

Resistance is usually measured with a *digital multimeter* (DMM), but it can also be measured using a *volt-ohm-milliammeter* (VOM). VOMs are not as accurate as DMMs in most applications and are rarely used today. However, it is still a good idea to be able to use a VOM properly. In this exercise, you will measure resistor values using both types of meters.

LAB PREPARATION

Review:

- Sections 2.2 and 2.3 of *Electronics Technology Fundamentals.*
- Appendix B of this manual (which covers the use of the DMM and VOM to measure resistance).

MATERIALS

1	DMM
1	VOM
10	assorted resistors (*Note:* No more than two resistors should have the same color multiplier band.)
1	1 MΩ resistor

PROCEDURE

1. Record the color bands of the 10 assorted resistors in Column 1 of Table 1.1.

TABLE 1.1   Measured and Calculated Resistor Values (DMM)

Color Bands	Nominal Value	Tolerance (%)	Value Range	Measured Value	Variation (%)

*Use the following abbreviations to indicate the color band values: Black (Bk), Brown (Bn), Red (Rd), Orange (Or), Yellow (Yw), Green (Gn), Blue (Bl), Violet (Vt), Gray (Gy), White (Wh)

2. Using the color code, determine the rated (nominal) value of each resistor, and record this value in Column 2 of the table.
3. Determine the tolerance of each resistor based on the color of its tolerance band, and record this value in Column 3 of the table.
4. Using the rated value and the tolerance of each resistor, calculate its expected maximum and minimum values. Record these values in Column 4 of the table.
5. Measure the value of each resistor using the DMM, and record this value in Column 5 of the table. Remember: *only one of your hands should come in contact with a resistor lead when measuring resistance,* as illustrated in Figure 1.1.

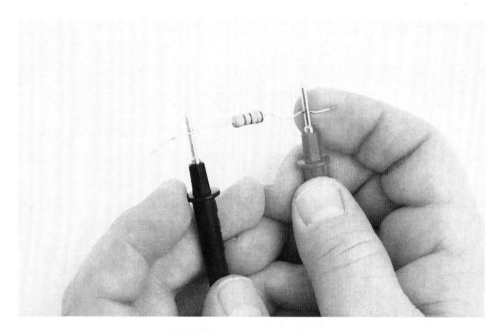

FIGURE 1.1 Measuring resistance.

6. The percent of variation between a nominal (rated) value and a measured value is calculated as follows:

$$\% \text{ Variation} = \frac{|\text{Nominal value} - \text{Measured value}|}{\text{Nominal value}} \times 100$$

Calculate the percent of variation for each resistor. Record your calculated values in Column 6 of the table.
7. Repeat Steps 1 through 6 using an analog VOM in place of the DMM. Remember:
   • Use the proper range.
   • Zero the meter each time you change ranges.
   • Avoid *parallax* errors (as described in Appendix B).
   Record your results in Table 1.2.

TABLE 1.2   Measured and Calculated Resistor Values (VOM)

Color Bands	Nominal Value	Tolerance (%)	Value Range	Measured Value	Variation (%)

*Use the following abbreviations to indicate the color band values: Black (Bk), Brown (Bn), Red (Rd), Orange (Or), Yellow (Yw), Green (Gn), Blue (Bl), Violet (Vt), Gray (Gy), White (Wh)

8. Using the DMM, measure the value of your 1 MΩ resistor.

$$R = \underline{\hspace{2cm}}$$

9. Repeat Step 8. This time, however, make sure that both your hands are touching the metal leads of the resistor (one hand on either lead). Record your measurement.

$$R = \underline{\hspace{2cm}}$$

10. Select a resistor with a red multiplier band and record its rated value. Then repeat Step 8.

$R = \underline{\hspace{2cm}}$ (rated) $R = \underline{\hspace{2cm}}$ (measured)

11. Using the same resistor, repeat Step 9.

$$R = \underline{\hspace{2cm}}$$

QUESTIONS & PROBLEMS

1. Assume that a 47 kΩ resistor and a 22 kΩ resistor each vary by 0.5% from their nominal values. Which of these resistors will vary by the largest actual value? Explain your answer.

2. A 330 kΩ resistor has a measured value of 338 kΩ. A 1.1 kΩ resistor has a measured value of 1.066 kΩ. Which component has the greatest variation percentage between its rated and measured values?

3. How much did your measurement in Step 9 vary from your measurement in Step 8? Which measurement was more accurate? Explain your answer.

4. Assuming that the value in Step 8 is more accurate than the value obtained in Step 9, calculate the percent of variation introduced by holding the component as you did in Step 9. (Use the value from Step 8 in the denominator of the variation equation.)

5. Assuming that the value in Step 10 is more accurate than the value in Step 11, calculate the percent of variation introduced by holding the component as you did in Step 11. (Use the value from Step 10 in the denominator of the variation equation.)

6. Compare your DMM and VOM resistance measurements. Based on your experience in this exercise, which meter would you prefer to use? Explain the reason for your choice.

SIMULATION EXERCISE

Discussion

In this simulation exercise you will use a multimeter to measure 10 resistors. Based upon your measurements you will then determine the closest standard value for each resistor, and its color code.

Procedure

1. Open file Ex1.1 from the Electronics Technology Fundamentals companion web site (www.prenhall.com/paynter). As you can see, there are ten resistors and a multimeter on your workspace. The resistor values are hidden.
2. Open the multimeter by double-clicking on it. Turn the meter on, run the simulation, and read the measured value of R_1. Record this value in Table 1.3.

TABLE 1.3   Measured and Nominal Resistor Values and Color Codes

	Measured Value	Nominal Value	Color Code
R_1			
R_2			
R_3			
R_4			
R_5			
R_6			
R_7			
R_8			
R_9			
R_{10}			

3. Using Table 2.3 on page 49 of the text, determine the closest standard resistor value that corresponds to the measured value of R_1. Record this value in Table 1.3.
4. Finally, determine the color bands for R_1.
5. Repeat Steps 1 through 4 for the remaining 9 resistors and complete Table 1.3.
6. Double-click on R_{10} and choose the Fault tab. Now choose "Open" and terminals 1 and 2. Measure the resistance of R_{10} faulted open and record this value below.

$$R_{10(open)} = \underline{\hspace{2cm}}$$

7. Repeat Step 6, but choose the "Short" fault.

$$R_{10(shorted)} = \underline{\hspace{2cm}}$$

Question

1. Refer to Steps 6 and 7. Were these measurements consistent with your understanding of what a short and open are? If a resistor that was much higher or lower in value than R_{10} was chosen, do you think that the measurements for these two faults would have been any different? Explain your answer.

Exercise 2

Potentiometers

OBJECTIVES

After completing this exercise, you should be able to:

- Determine the rated value of a potentiometer.
- Demonstrate the resistance relationships among the three terminals of a potentiometer.
- Describe the concept of *taper* as it applies to potentiometers.

DISCUSSION

A *potentiometer* is a three-terminal resistor whose value can be adjusted (within specified limits) by rotating a control shaft. Figure 2.1 shows the pictorial and schematic representations of a potentiometer. Note that the three terminals are labeled as A, B, and C, with terminal B always being the center terminal. The center terminal is often referred to as the "*wiper*". The resistance between terminals A and C (R_{AC}) is fixed and does not change as the potentiometer is adjusted. Note that R_{AC} is the *nominal* (rated) value of the potentiometer.

Figure 2.2 shows a simplified representation of the internal structure of a potentiometer. The resistance between terminal B and the other two terminals is adjusted by rotating the control shaft. The rate of change in potentiometer resistance as the control shaft rotates is called the *taper* of the potentiometer. There are two types of taper: linear and nonlinear. The resistance of a linear-taper potentiometer varies at a fixed rate as the control shaft is turned. The resistance of a nonlinear-taper potentiometer varies at increasing rates as the control shaft is turned from one extreme to the other. Note that the term *audio-taper* is often used to describe a nonlinear taper.

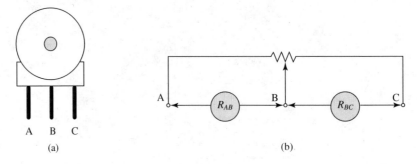

FIGURE 2.1   Pictorial and schematic representations of a potentiometer.

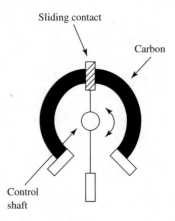

FIGURE 2.2   Internal representation of a potentiometer.

LAB PREPARATION

Review:

- Section 2.4 of *Electronics Technology Fundamentals.*
- The guidelines for measuring resistance in Appendix B of this manual.

MATERIALS

1   DMM
1   10 kΩ linear-taper potentiometer
1   10 kΩ audio-taper potentiometer
1   10 kΩ, 10-turn potentiometer

PROCEDURE

1. Select the linear-taper potentiometer, and refer to Figure 2.3. Rotate the control shaft fully counter-clockwise. Mark this point on the body of the pot, as shown in Figure 2.3. Then rotate the shaft fully clockwise. Mark this point. As closely as possible, estimate the ¼, ½, and ¾ rotation positions (from left to right).

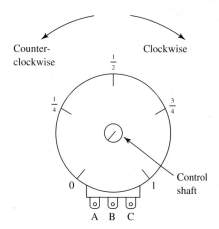

FIGURE 2.3   Positions and electrical connections.

2. Rotate the control shaft fully counter-clockwise, and measure the resistance between the outer terminals (R_{AC}). Enter the measured value in Table 2.1.

TABLE 2.1   Linear Resistance Measurements

Shaft Position	R_{AC}	R_{AB}	R_{BC}
Fully counter-clockwise			
¼ turn clockwise			
½ turn clockwise			
¾ turn clockwise			
Fully clockwise			

3. Measure the resistance between the center terminal and each outer terminal (R_{AB} and R_{BC}), and record the measurements in Table 2.1.
4. Repeat Steps 2 and 3 for the ¼, ½, ¾, and fully clockwise positions.
5. Using your R_{AB} values from Table 2.1, plot the resistance-versus-rotation curve for this potentiometer in Figure 2.4. Represent rotation on the *x*-axis and resistance (the dependent variable) on the *y*-axis.

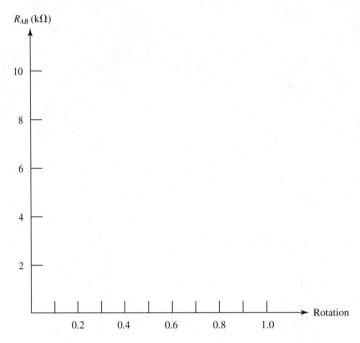

FIGURE 2.4 Graph of resistance versus rotation.

6. Repeat Steps 1 through 4 for the audio-taper potentiometer, and record your measurements in Table 2.2.

TABLE 2.2   Nonlinear Resistance Measurements

Shaft Position	R_{AC}	R_{AB}	R_{BC}
Fully counter-clockwise			
¼ turn clockwise			
½ turn clockwise			
¾ turn clockwise			
Fully clockwise			

7. Plot the resistance-versus-rotation curve for the audio-taper potentiometer on the same graph as the linear-taper potentiometer.

10-Turn Potentiometer

8. Select the precision 10-turn potentiometer. Measure R_{AC}, and record this value in Table 2.3.

TABLE 2.3   10-Turn Potentiometer Readings

Rotations	R_{AC}	R_{AB}	R_{BC}
0			
1			
2			
3			
4			
5			
6			
7			
8			
9			
10			

9. Rotate the control shaft fully counter-clockwise. Measure the values of R_{AB} and R_{BC}. Record your measurements in the "0 Rotations" row of Table 2.3.
10. Repeat Step 9 for 1 through 10 rotations of the control shaft, and record the measurements in the table.
11. Using your results for R_{AB} from Table 2.3, plot the resistance-versus-rotation graph for this potentiometer in Figure 2.5.

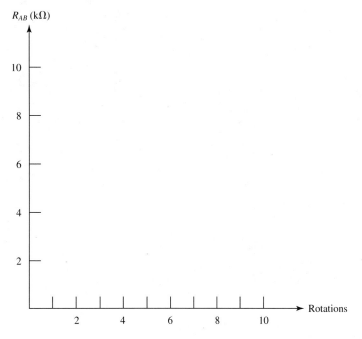

FIGURE 2.5   Graph of resistance-versus-rotation for the 10-turn potentiometer.

QUESTIONS & PROBLEMS

1. Citing examples from Tables 2.1 through 2.3, explain the relationship among R_{AC}, R_{AB}, and R_{BC}.

2. Based on your results in Tables 2.1 and 2.2 and the plots in Figure 2.4, explain the difference between a linear taper and a nonlinear (audio) taper.

3. Based on your results in Table 2.3 and the plot in Figure 2.5, did the 10-turn precision potentiometer have a linear or a nonlinear taper? Explain your reasoning.

4. The *resolution* of a potentiometer is the change in resistance per degree of rotation, found as

$$\text{Resolution} = \frac{\Delta R}{^\circ \text{Rotation}}$$

Calculate the resolution of your standard linear potentiometer and your 10-turn potentiometer. How do the results compare?

Resolution = _____ (standard linear potentiometer)

Resolution = _____ (10-turn potentiometer)

SIMULATION EXERCISE

Procedure

1. Open file Ex2.1 from the Electronics Technology Fundamentals companion web site (www.prenhall.com/paynter). As you can see there is a 10 kΩ potentiometer on your workspace with its terminals labeled as A, B and C. Terminal B is the wiper. There is also a multimeter set to function as an ohmmeter on your workspace.
2. Note that the potentiometer is set at 50%. This is equivalent to the ½-turn setting in Figure 2.3. Run the simulation and use the ohmmeter to verify that $R_{AC} = 10$ kΩ and $R_{AB} = R_{BC} = 5$ kΩ.
3. Change the setting of the potentiometer to the values shown in Table 2.4. Note that typing "a" increases the potentiometer percentage and typing "shift a" decreases the percentage. Measure R_{AB}, R_{BC}, and R_{AC} for each setting in the Table 2.4 and record your measurements in the table.

TABLE 2.4 Simulated Resistance Readings for a 10 kΩ Potentiometer

% Setting	R_{AB}	R_{BC}	R_{AC}
0%			
20%			
40%			
60%			
80%			
100%			

4. Return the wiper arm setting to the 50% position, then use the fault feature to open each potentiometer lead individually. Measure the resulting faulted component values with the ohmmeter, and record these values in Table 2.5.

TABLE 2.5   Simulation Results for Faulted Potentiometer Terminals

Fault Condition	R_{AB}	R_{BC}	R_{AC}
Terminal A open			
Terminal B open			
Terminal C open			

Questions

1. Do your results from Table 2.4 support the relationship among R_{AB}, R_{BC}, and R_{AC} that you have learned? Explain your answer.

2. Is the potentiometer used in this simulation a linear or audio-taper potentiometer? Explain your answer.

3. Refer to your results in Table 2.5. What effect does an open end-terminal have on the resistance between the wiper and the other end-terminal? Explain your answer.

Exercise 3

Ohm's Law

OBJECTIVES

After completing this exercise, you should be able to:

- Measure voltage using a multimeter.
- Measure current using a multimeter.
- Demonstrate the relationship among current, voltage, and resistance.

DISCUSSION

Appearances can be very deceiving. For example, $V = I \times R$ does not look very impressive as equations go, but don't be fooled. This equation is a form of Ohm's law, a very powerful tool that you will use every day of your professional life. It is a principle that must be grasped totally, because it forms the basis for all other laws and theorems in electronics. As such, its importance cannot be overstated.

Ohm's law describes the relationship among current, voltage, and resistance. In this exercise, you will examine these relationships and build your first circuit. Even though this circuit will be quite simple, now is the time to develop proper laboratory techniques. Be sure to follow all the procedure steps carefully and accurately.

LAB PREPARATION

Review:

- Sections 3.1 through 3.3 of *Electronics Technology Fundamentals*.
- The circuit construction guidelines in Appendix A.
- The guidelines for measuring current and voltage in Appendix B.

MATERIALS

2   DMMs
1   variable dc power supply
1   protoboard
1   10 kΩ, 2 W potentiometer
4   resistors: 1 kΩ, 2.2 kΩ, 4.7 kΩ, and one randomly chosen resistor that is greater than 1 kΩ (All resistors are ¼ W.)

PROCEDURE

1. Measure and record the values of the resistors listed in Table 3.1.

TABLE 3.1

Nominal	Measured
1 kΩ	
2.2 kΩ	
4.7 kΩ	

2. Using Ohm's law, calculate the current through the circuit in Figure 3.1, then repeat this procedure for the other two resistors. Record your calculated current values in Table 3.2.

Note: Always use measured component values (when possible) in your circuit calculations.

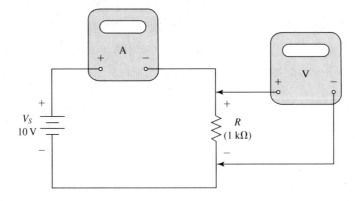

FIGURE 3.1

TABLE 3.2   Current Measurements

V (Measured)	R (Measured)	$I = \dfrac{V}{R}$	I (Measured)	Variation (%)
10 V				
10 V				
10 V				
6 V				
6 V				
6 V				

3. Change V_S to 6 V, and repeat Step 2.
4. Construct the circuit shown in Figure 3.1. Make certain that the dc power supply is off and set to its lowest value before you start construction.
5. When the circuit is completed, turn on the dc power supply. Slowly increase power supply output until you read 10 V on the voltmeter.
6. Record the ammeter current reading in Table 3.2 and then turn the dc power supply off.
7. Determine the percent of variation between the calculated and measured value of current. Record this value in Table 3.2.
8. Repeat Steps 5 through 7 for the 2.2 kΩ and 4.7 kΩ resistors. Record your results in Table 3.2.
9. Repeat Steps 5 through 8 for $V_S = 6$ V.
10. Replace the 1 kΩ resistor shown in Figure 3.1 with your random-value resistor. Set the power supply to 10 V. Measure the circuit current and record its value in Table 3.3.

TABLE 3.3   Data for Steps 10, 11, and 12

V_S	I (Measured)	$R = \dfrac{V}{I}$	R (Measured)	Variation (%)
10 V				

11. Turn off the dc power supply. Use Ohm's law to calculate the circuit resistance using your measured values of voltage and current. Enter the calculated resistance in Table 3.3.
12. Remove the resistor and measure its value. Enter this value in Table 3.3. Determine the percent of variation between the calculated and measured values of the resistor.
13. Construct the circuit shown in Figure 3.2. Make certain that the potentiometer is set to its maximum resistance.

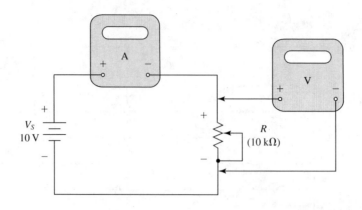

Note: The potentiometer should be set initially to its maximum resistance.

FIGURE 3.2

14. Turn on the dc power supply, and adjust its output to 10 V. Adjust the potentiometer until you get a current reading of 2.5 mA.
15. Turn off the power supply, and remove the potentiometer. Measure its adjusted value, and record your result in Table 3.4.
16. Repeat Steps 14 and 15 for current values of 5 mA, 7.5 mA, and 10 mA.
17. Using the results from Table 3.4, plot the current-versus-resistance curve in Figure 3.3.

TABLE 3.4

V_S	I (Measured)	R (Measured)
10 V	2.5 mA	
10 V	5 mA	
10 V	7.5 mA	
10 V	10 mA	

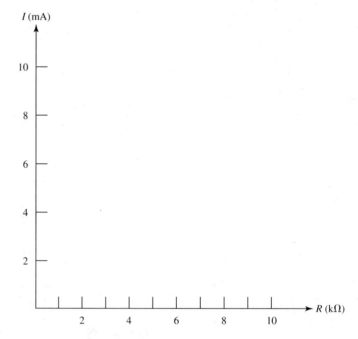

FIGURE 3.3 Graph of the data from Table 3.4.

1. Based on your results in Table 3.2, explain what happens to current when voltage is fixed and circuit resistance increases.

2. Explain why measured component values are preferred to nominal values when performing circuit calculations.

3. Assume that you do not have a voltmeter but do have a functioning ammeter. Explain how you could use Ohm's law to estimate the value of V_S.

4. Refer to the graph in Figure 3.3. Describe the rate at which current changes with a change in resistance when voltage is held constant.

5. In Tables 3.2 and 3.3, you were asked to determine the percent of variation between your measured and calculated values. Comment on any variations and why you think they occurred.

SIMULATION EXERCISE

Discussion

One of the initial challenges for any student is learning how to quickly and correctly identify the resistance values associated with *open* and *short* circuits. An open circuit *reduces the amount of current,* which implies an increase in the circuit resistance. A short circuit *increases the amount of current,* which commonly activates a safety device such as a circuit breaker or fuse. This exercise allows you to experiment with open circuits, short circuits, and fuses in a safe and economical manner.

Procedure

1. Open file Ex3.1 from the Electronics Technology Fundamentals companion web site (www.prenhall.com/paynter). There are two voltage indicators and one current indicator on your workspace.

   Note: All voltage indicators used in these simulation exercises have their internal resistance set to 10 MΩ.

2. Using the values shown in the circuit, calculate the circuit current.

   $$I = \text{_____}$$

3. Run the simulation, and record the ammeter reading in the first row of Table 3.5. How does the ammeter reading compare with the current value that you calculated in Step 2?

TABLE 3.5

Circuit Condition	I	V_R	V_F	Fuse Condition
Normal operation				Good
Resistor open				
Resistor shorted				

4. Measure the voltage across R (which we will designate as V_R). Record this reading in the first row of Table 3.5.

5. Record the voltage across the fuse (which we will designate as V_F) in the first row of Table 3.5. Then, using V_F and I, calculate the resistance of the fuse.

$$R_F = \underline{\hspace{3cm}}$$

6. Fault the resistor open as follows:
    a. Double-click on the resistor symbol.
    b. Select Fault.
    c. Select Open, and check either terminal 1 or terminal 2.

7. Run the simulation, and complete line 2 of Table 3.5.

8. Return to the resistor fault screen, and change the fault from open to short. (Be sure to check both terminals 1 and 2.)

> *Safety Note:* 1 A is a dangerously high value of current. Simulations allow you to test high-current circuits in a safe manner.

9. Run the simulation, and complete line 3 of Table 3.5. *Note:* The fuse is replaced each time you restart the simulation.

Questions

1. Which reading (or combination of readings) in the second row of Table 3.5 indicates the existence of a problem? Explain your answer.

2. Which reading (or combination of readings) in the third row of Table 3.5 indicates the existence of a problem? Explain your answer.

3. What is the resistance of an open component?

4. What is the resistance of a shorted component?

5. What kind of problem is indicated by a blown fuse?

Exercise 4

Power

OBJECTIVES

After completing this exercise, you should be able to:

- Demonstrate the relationship between power and heat.
- Calculate component power dissipation when two of the three basic circuit values are known.
- Demonstrate the relationship between component power dissipation and total circuit power.
- Demonstrate the relationship between power and current.

DISCUSSION

The rules governing power in electrical circuits can be viewed as extensions of Ohm's law. In fact, the basic power equations are simple variations on the Ohm's law equations.

Power and heat are closely related. When a resistor uses electrical energy, it produces heat. In this exercise, you will investigate the relationship between power and heat.

As you know, the voltage across a fixed resistance varies in direct proportion to the current through that resistance. In contrast, the power that is dissipated by a fixed resistance varies in direct proportion to *the square of the current*. The relationship between power and current will be demonstrated in this exercise.

LAB PREPARATION

Review:

- Section 3.4 of *Electronics Technology Fundamentals*.
- The circuit construction guidelines in Appendix A.
- The techniques for measuring voltage and current in Appendix B.

MATERIALS

2   DMMs
1   variable dc power supply
1   protoboard
2   resistors: 100 Ω and 470 Ω (¼ W)

PROCEDURE

1. Construct the circuit shown in Figure 4.1. Measure and record the circuit current in Table 4.1.

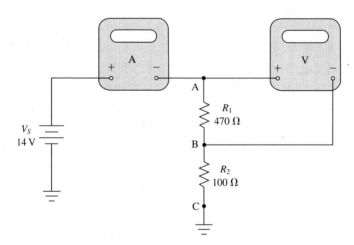

FIGURE 4.1

TABLE 4.1

Measured Values				Calculated Values			
V_S	I	V_{R1}	V_{R2}	I^2R_1	I^2R_2	V_{R1}^2/R_1	V_{R2}^2/R_2
14 V							

2. Measure the voltages across $R_1(V_{R1})$ and $R_2(V_{R2})$. Record your results in Table 4.1.
3. After the circuit has been running for 1 to 2 minutes, *carefully* touch the body of each resistor. (*Note:* The components may be hot, so touch them briefly and lightly. Do *not* touch the metal leads.) Which resistor (if either) seems to be warmer?

4. Turn off the dc power supply. Using the appropriate power equations and your measured values, complete Table 4.1.
5. Using the measured values of V_S and I from Table 4.1, calculate the total circuit power (P_T).

$$P_T = \underline{\hspace{3cm}}$$

6. Construct the circuit shown in Figure 4.2. Adjust the power supply output to its minimum setting.

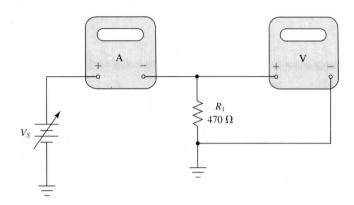

FIGURE 4.2

7. Turn on the dc power supply, and increase its output until you get a current reading of 1 mA. Measure V_{R1}, and record its value in Table 4.2.

TABLE 4.2

	Measured Values		Calculated Values		
I	V_{R1}	$I \times V_{R1}$	$I^2 \times R_1$	$\dfrac{V_{R1}^2}{R_1}$
1 mA				
2 mA				
3 mA				
4 mA				
5 mA				

8. Repeat Step 7 for current values of 2 mA, 3 mA, 4 mA, and 5 mA.
9. For each current level, calculate the power dissipated by the resistor using all three power formulas.

10. Using your results from Table 4.2, plot a power-versus-current curve in Figure 4.3.

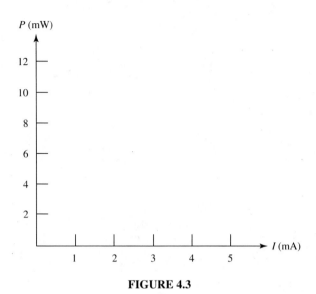

FIGURE 4.3

QUESTIONS & PROBLEMS

1. Refer to Step 3. Which resistor seemed to be warmer? Without using any calculations, explain why you think this was the case.

2. Compare the calculated power values in Table 4.1. Are there any significant differences between the calculated values in each row? How would you explain any differences between the calculated values?

3. Compare the calculated power values in Table 4.1 with the value of P_T found in Step 5. Describe the relationship between the component power values and the total circuit power.

4. Refer to your results in Table 4.2. Did the results for all three methods of calculating power agree? If not, what do you believe caused any discrepancies?

5. Refer to the graph in Figure 4.3. Describe how power dissipation varies with changes in current.

SIMULATION EXERCISE

Procedure

1. Open file Ex4.1 from the Electronics Technology Fundamentals companion web site (www.prenhall.com/paynter). This circuit is very similar to the circuit in Figure 4.1, but the value of V_S has been increased to 20 V.
2. Run the simulation and measure the values of I_T, V_{R1}, and V_{R2}. Record these values in Table 4.3.
3. Complete Table 4.3 by calculating the power dissipated by each resistor as shown in the table.

TABLE 4.3 Simulation Results for Normal Circuit Operation

Measured Values				Calculated Values			
V_S	I	V_{R1}	V_{R2}	I^2R_1	I^2R_2	V_{R1}^2/R_1	V_{R2}^2/R_2
20 V							

4. Modify your simulation circuit as follows:
   - Double-click the 100 Ω resistor symbol.
   - Click the Fault tab.
   - Select Short as the fault, and place checks in both terminals 1 and 2.
   - Click OK.
5. Run the simulation, and complete Table 4.4.

TABLE 4.4   Simulation Results for Circuit Operation with R_2 Shorted

Measured Values				Calculated Values			
V_S	I	V_{R1}	V_{R2}	I^2R_1	I^2R_2	V_{R1}^2/R_1	V_{R2}^2/R_2
20 V							

Questions

1. Compare the V_{R1} measurements from Table 4.3 and 4.4. Explain why this voltage measurement changed in the way that it did.

2. Refer to your results in Table 4.3. First, use the values of V_S and I_T to calculate the total circuit power and record this value below. Then add the values of P_{R1} and P_{R2} and record this value below. Explain the relationship between these two calculated values.

   $P_T = V_S \times I_T =$ _____ $P_{R1} + P_{R2} = I^2R_1 + I^2R_2 =$ _____

3. Suppose that R_2 was open instead of shorted. What would happen to the value of P_T? Support your answer.

Exercise 5

Series Circuits and Kirchhoff's Voltage Law

OBJECTIVES

After completing this exercise, you should be able to:

- Demonstrate the current characteristics of series circuits.
- Calculate the total resistance in a series circuit.
- Calculate the total current through a series circuit.
- Apply Kirchhoff's voltage law to a basic series circuit.
- Solve for an unknown resistor using Kirchhoff's voltage law and Ohm's law.
- Explain the effect of a shorted or an open resistor on series circuit operation.

DISCUSSION

There are two basic types of circuits: *series* circuits and *parallel* circuits. In the next two exercises, you will concentrate on series circuits.

The series circuit can be defined as a circuit that has only one current path. It may contain any number of components, but the same current passes through all of them. As such, the current in a series circuit is equal at all points in the circuit.

Simply stated, *Kirchhoff's voltage law* says that the sum of the resistor voltages in a series circuit must equal the applied voltage. Since current is equal through all the components in a series circuit, the sum of the resistor power dissipation values must also equal the total circuit power.

The total resistance in a series circuit equals the sum of the resistor values. If all but one of the resistor values is known and you know the value of circuit current, Ohm's law can be used to calculate the value of the unknown resistor.

In this exercise, you will look at all the concepts just described. You will also see what happens to a series circuit if one of the components fails, either as an open or as a short.

LAB PREPARATION

Review:

- Sections 4.1 and 4.2 of *Electronics Technology Fundamentals*.
- The techniques for measuring voltage and current in Appendix B.

MATERIALS

2   DMMs
1   variable dc power supply
1   protoboard
3   resistors: 1.2 kΩ, 2.2 kΩ, and 3.3 kΩ (all resistors are ¼ W)

PROCEDURE

1. Measure the values of the resistors that you will use to construct the circuit shown in Figure 5.1. Enter their measured values in Table 5.1.

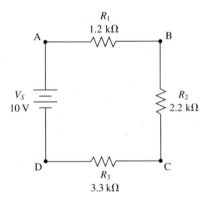

FIGURE 5.1

TABLE 5.1

Component	Nominal Value	Measured Value
R_1	1.2 kΩ	
R_2	2.2 kΩ	
R_3	3.3 kΩ	

2. Construct the circuit shown in Figure 5.1 with an ammeter inserted at point A. Apply power, and measure the current at this point (which we will designate as I_A). Record your measurement in Table 5.2.

TABLE 5.2

V_S	I_A	I_B	I_C	I_D
10 V				

3. Disconnect power from the circuit, and move the ammeter to point B. Restore power, and record the measured current in Table 5.2 as I_B.
4. Repeat Step 3 for points C and D in the circuit, and record your results in Table 5.2.
5. Disconnect power from the circuit, and remove the ammeter.
6. Restore power to the circuit, and measure the voltage across R_1. Record your measurement in Table 5.3.

TABLE 5.3

V_{R1}	I_{R1}	V_{R2}	I_{R2}	V_{R3}	I_{R3}

7. Use Ohm's law to calculate the current through R_1, and record your calculation in Table 5.3.
8. Repeat Steps 6 and 7 for the other resistors in the circuit. After completing your measurements, disconnect power from the circuit.
9. Based on your results in Table 5.3, determine the sum of the three resistor voltages in the circuit.

$$V_{R1} + V_{R2} + V_{R3} = \text{_____}$$

10. Using the circuit values of V_S and I from Table 5.2, calculate the total circuit power (P_T). Enter your calculation in Table 5.4.

TABLE 5.4

P_T	P_{R1}	P_{R2}	P_{R3}

11. Using the appropriate values from Table 5.3, calculate the power dissipated by each resistor. Enter your results in Table 5.4.

QUESTIONS & PROBLEMS

1. Refer to Table 5.2. What do these results tell you about the current through a series circuit?

2. Did your result in Step 9 substantiate Kirchhoff's voltage law? Explain your answer.

3. Refer to Table 5.4. Based on what you know about Kirchhoff's voltage law and your results, explain the relationship between individual component power dissipation and total circuit power in a series circuit.

4. Suppose the color code of a resistor is unreadable. Based on what you have learned in this exercise, explain how you could use Kirchhoff's voltage law and Ohm's law to find the value of this unknown resistor in a series circuit.

SIMULATION EXERCISE

Procedure

1. Open file Ex5.1 from the Electronics Technology Fundamentals companion web site (www.prenhall.com/paynter). This circuit is the same as Figure 5.1, but it has been heavily instrumented. Run the simulation and verify that:
    - The current readings are the same at all points in the circuit.
    - The sum of the resistor voltages is equal to the source voltage.

2. Modify your simulation circuit as follows:
   - Double-click the symbol for R_2.
   - Click the Fault tab.
   - Select Open as the fault, and check either terminal 1 or terminal 2.
   - Click OK.
3. Run the simulation with R_2 faulted open, and answer the following questions:
   a. What are the ammeter indications with the fault present? Why don't they read 0 A? (*Hint:* Remember that voltmeters aren't perfect.)

   b. In your opinion, are the current readings significant? Explain your reasoning.

   c. What are the voltmeter readings with the fault present? Why are V_{R1} and V_{R3} not equal to 0 V? Explain the V_{R2} reading.

   d. In your opinion, are the V_{R1} and V_{R3} readings significant? Explain your reasoning.

4. Modify your simulation circuit as follows:
   - Double-click the symbol for R_2.
   - Click the Fault tab.
   - Select Short as the fault, and check both terminals 1 and 2.
   - Click OK.

5. Run the simulation with R_2 faulted as a shorted component, and answer the following questions:

    a. How does the circuit current compare to its "normal" value?

    b. Do the voltmeter readings indicate which resistor is shorted? Explain your answer.

    c. What is the relationship between the sum of V_{R1} and V_{R3} and the source voltage? How would you explain this relationship?

6. Modify your simulation circuit as follows:
    *   Double-click the symbol for R_2.
    *   Click the Fault tab.
    *   Select Leakage as the fault, and set the leakage value to $\frac{1}{10}$ the normal value of the resistor.
    *   Click OK.
    This procedure takes R_2 out of tolerance (rather than shorting it completely).

7. Run the simulation with R_2 out of tolerance, and answer the following questions:

   a. How does the circuit current compare to its "normal" value?

   b. Do the voltmeter readings indicate which resistor is out of tolerance? Explain your answer.

   c. Explain the changes in the values of I_T, V_{R1}, and V_{R3}.

Questions

1. Any number of current and voltage indicators can be placed in a simulation circuit, allowing you to take all the circuit readings at once. At the same time, economic considerations typically limit technicians to one or two meters. Explain the effects that a large number of meters can have on the accuracy of simulation results. How can these effects be minimized?

2. The first simulation was run with the circuit operating normally. Explain how knowing the "normal" circuit values can help to determine a problem's location and type.

Exercise 6

The Voltage-Divider Relationship and Source Resistance

OBJECTIVES

After completing this exercise, you should be able to:

- Explain the distribution of source voltage and power in a series circuit.
- Demonstrate the concept of proportionality as it applies to voltage dividers.

DISCUSSION

The *voltage-divider rule* states that the ratio of component voltage to applied voltage equals the ratio of component resistance to total circuit resistance. For example, if a resistor provides 25% of the total circuit resistance, then 25% of the applied voltage is developed across that component. This relationship holds true regardless of the applied voltage or the actual value of the resistors. (It's all about proportionality.)

LAB PREPARATION

Review Section 4.2 of *Electronics Technology Fundamentals*.

MATERIALS

1 DMM
1 variable dc power supply
1 protoboard
4 resistors: 1.2 kΩ, 2.2 kΩ, 3.3 kΩ, and one randomly selected resistor with a red multiplier band (All resistors are ¼ W.)

PROCEDURE

1. Measure the values of the three resistors you will use in the circuit shown in Figure 6.1. Enter the resistor values in Table 6.1.

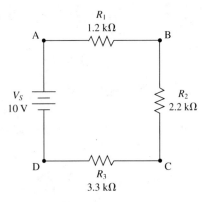

FIGURE 6.1

TABLE 6.1

Component	Nominal Value	Measured Value
R_1	1.2 kΩ	
R_2	2.2 kΩ	
R_3	3.3 kΩ	

2. Refer to the circuit shown in Figure 6.1. Use the voltage-divider equation to calculate the voltage across each resistor in the circuit. Be sure to use your measured component values.

$$V_{R1} = V_S \frac{R_1}{R_T} = \underline{\hspace{2cm}}$$

$$V_{R2} = V_S \frac{R_2}{R_T} = \underline{\hspace{2cm}}$$

$$V_{R3} = V_S \frac{R_3}{R_T} = \underline{\hspace{2cm}}$$

3. Construct the circuit shown in Figure 6.1. Power up and measure the voltage across each resistor in the circuit. Record your results in Table 6.2.

TABLE 6.2

V_S	V_{R1}	V_{R2}	V_{R3}
10 V			

4. Using your measured values from Tables 6.1 and 6.2, calculate the percentages identified in Table 6.3.

TABLE 6.3

$\dfrac{R_1}{R_T} \times 100$	$\dfrac{V_{R1}}{V_S} \times 100$	$\dfrac{R_2}{R_T} \times 100$	$\dfrac{V_{R2}}{V_S} \times 100$	$\dfrac{R_3}{R_T} \times 100$	$\dfrac{V_{R3}}{V_S} \times 100$
%	%	%	%	%	%

5. Refer back to Figure 6.1. Use the voltage-divider relationship to determine the voltage from point A to point C and from point B to point D.

$$V_{A-C} = V_S \frac{R_1 + R_2}{R_T} = \underline{\hspace{3cm}}$$

$$V_{B-D} = V_S \frac{R_2 + R_3}{R_T} = \underline{\hspace{3cm}}$$

6. Measure the voltages calculated in Step 5, and record your findings in Table 6.4.

TABLE 6.4

V_S	V_{A-C}	V_{B-D}
10 V		

7. Disconnect power from the circuit, and replace R_3 with your unknown resistor. Restore power, and measure the voltages across R_1 and R_3.

$$V_{R1} = \underline{\hspace{3cm}}$$

$$V_{R3} = \underline{\hspace{3cm}}$$

8. The ratio of R_1 to R_3 must equal the ratio of V_{R1} to V_{R3}. By formula,

$$\frac{R_3}{R_1} = \frac{V_{R3}}{V_{R1}}$$

Using the relationship below and your measured values of R_1, V_{R1}, and V_{R3}, calculate the value of R_3.

$$R_3 = R_1 \frac{V_{R3}}{V_{R1}} = \underline{\hspace{3cm}}$$

1. Refer to Steps 2 and 3. Did your results substantiate the voltage-divider rule? Explain your answer.

2. Explain how the results in Table 6.3 illustrate the concept of proportionality (as it applies to the voltages in a series circuit).

3. Explain how your results in Steps 5 and 6 demonstrate the voltage-divider rule.

4. Refer to Steps 7 and 8. Explain your results in terms of the voltage-divider relationship and the concept of proportionality.

Discussion

Any circuit that provides voltage to another circuit or load can be considered to be a voltage source. As your electronics education continues, you will learn that the relationship between the output resistance of a voltage source and the resistance of its load can have a profound effect on how efficient the voltage source is. This concept is explored in detail in Exercise 12.

An ideal voltage source would have no output resistance, but this is never the case in practice. This concept of *the resistance of a source as seen by its load* is so prevalent that it is known by a variety of names: source resistance, output resistance, and Thevenin resistance. Since source resistance cannot be measured directly with an ohmmeter, it is useful to know another way to determine its value.

The output voltage of a voltage source, with no load connected, is referred to as the no-load voltage (V_{NL}). In this simulation exercise you will learn how to use V_{NL}, V_L, R_L, and the laws of proportionality to determine source resistance.

Procedure

Simulators provide *ideal* voltage sources, meaning that $R_S = 0\,\Omega$. A *real* (or *practical*) voltage source has a value of $R_S > 0\,\Omega$. For this reason, we must modify the simulator voltage source, to more closely resemble a practical voltage source, as follows:

1. Open file Ex6.1 from the Electronics Technology Fundamentals companion web site (www.prenhall.com/paynter). This circuit is the same as Figure 6.1, but the source voltage has been increased to 12 V and an internal resistance of 100 Ω (labeled as R_S) has been added. R_S should be treated as part of the practical voltage source.
2. Run the simulation, and record the meter reading as V_L.

$$V_L = \underline{\hspace{3cm}}$$

3. Stop the simulation. Leaving the meter connected, remove the wire between point A and R_1.
4. Run the simulation again, and measure the voltage between points A and B. Record this value as V_{NL}.

$$V_{NL} = \underline{\hspace{3cm}}$$

You have now measured the two voltages needed to determine the source resistance.

5. Now, solve for the source resistance as if its value were unknown. Using the measured voltages and the load resistance, the value of the source resistance can be found as follows:

a. The total load resistance is found as

$$R_L = R_1 + R_2 + R_3 = \underline{\hspace{3cm}}$$

b. The load current is found using Ohm's law:

$$I_L = \frac{V_L}{R_L} = \underline{\hspace{3cm}}$$

c. The voltage across the source resistance (V_{RS}) is found using Kirchhoff's voltage law:

$$V_{RS} = V_{NL} - V_L = \underline{\hspace{3cm}}$$

d. Since this is a series circuit, the source current equals the load current. Therefore,

$$R_S = \frac{V_{RS}}{I_L} = \underline{\hspace{3cm}}$$

6. Step 5 can be simplified using the concept of proportionality. As you know,

$$\frac{V_{RS}}{V_L} = \frac{R_S}{R_L}$$

This relationship can be transposed to:

$$R_S = R_L \left(\frac{V_{NL} - V_L}{V_L} \right)$$

Using the above equation and your measured values, calculate R_S.

$$R_S = \underline{\hspace{3cm}}$$

Questions

1. Comment on the following statement: *It is not sufficient to know Ohm's law and Kirchhoff's voltage law. You must also know how to apply them.*

2. Identify the simulation steps that demonstrate an application of Ohm's law and/or Kirchhoff's voltage law.

3. Using values from this exercise, demonstrate how the second equation in Step 6 could be modified to solve for R_L when the value of R_S is known.

Exercise 7

Parallel Circuits

OBJECTIVES

After completing this exercise, you should be able to:

- Measure the total resistance of a parallel circuit.
- Measure the total current through a parallel circuit.
- Demonstrate the power characteristics of a parallel circuit.

DISCUSSION

In Exercise 5, we stated that there are two basic types of circuits: series circuits and parallel circuits. In many ways, a parallel circuit is the direct opposite of a series circuit. For example:

- In a series circuit, current is the same through all components, and voltage is divided among them.
- In a parallel circuit, voltage is the same across all branches, and current is divided among them.

The resistance relationships in a parallel circuit are also very different from those in a series circuit. When a resistor is added to a series circuit, the total circuit resistance increases. When you add a resistor to a parallel circuit however, the total circuit resistance always *decreases.* There are three ways of solving for R_T in a parallel circuit. One is a general-applications equation; the other two are used in specific circumstances.

The one characteristic that series and parallel circuits have in common is that total circuit power equals the sum of the component power values. In this exercise, you will look at all of these concepts.

LAB PREPARATION

Review Sections 5.1 and 5.2 of *Electronics Technology Fundamentals.*

MATERIALS

2   DMMs
1   variable dc power supply
1   protoboard
6   resistors: 1 kΩ (3), 2.2 kΩ, 3.3 kΩ, and 4.7 kΩ

PROCEDURE

1. Measure the value of each resistor listed in Table 7.1, and enter your measured values in the table.

TABLE 7.1

Component	Nominal Value	Measured Value
R_1	2.2 kΩ	
R_2	3.3 kΩ	
R_3	4.7 kΩ	

TABLE 7.2

Approach	R_T
Reciprocal	
Product-over-sum	

2. Refer to the circuit shown in Figure 7.1. Use both the product-over-sum and the reciprocal formulas to solve for R_T. Enter your results in Table 7.2.

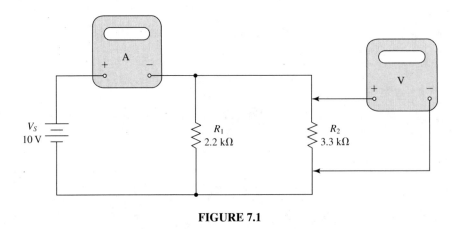

FIGURE 7.1

3. Construct the circuit shown in Figure 7.1. Apply power, and measure the current and voltage values listed in Table 7.3. Then, calculate the total circuit resistance using the measured values of I_T and V_S.

TABLE 7.3

I_T	V_S	V_{R1}	V_{R2}	R_T

4. Using your results from Step 3, make the necessary calculations to complete Table 7.4.

TABLE 7.4

$V_S \times I_T$	P_{R1}	P_{R2}	$P_{R1} + P_{R2}$

5. Construct the circuit shown in Figure 7.2. (*Note:* You are simply adding R_3 to the circuit shown in Figure 7.1.)

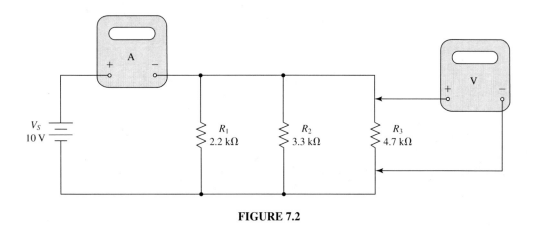

FIGURE 7.2

6. Calculate the total circuit resistance for this circuit.

$$R_T = \underline{\hspace{2cm}}$$

7. Apply power, and measure the current and voltage values listed in Table 7.5. Then, calculate the total circuit resistance using the measured values of I_T and V_S.

TABLE 7.5

I_T	V_S	V_{R1}	V_{R2}	V_{R3}	R_T

8. Repeat Step 4 for this circuit. Record your results in Table 7.6.

TABLE 7.6

$V_S \times I_T$	P_{R1}	P_{R2}	P_{R3}	$P_{R1} + P_{R2} + P_{R3}$

9. Construct the circuit shown in Figure 7.3. Apply power, and measure the total circuit current. Use this value to calculate the total circuit resistance. Then, use the equal-value resistor equation to calculate the value of R_T. Record your results in Table 7.7.

TABLE 7.7

Approach	R_T
Ohm's law	
Equal-value resistor equation	

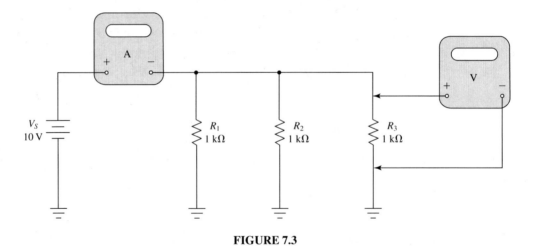

FIGURE 7.3

QUESTIONS & PROBLEMS

1. Refer to Step 2. Did the two approaches for calculating R_T provide the same result? Which method do you prefer? Why?

2. Did your calculated results for R_T in Step 3 agree with your calculations in Step 2? Comment on any discrepancies.

3. Compare the values of I_T and R_T in Step 3 with those in Step 7. What effect did adding R_3 to the circuit have on these values?

4. Compare the value of P_T in Step 4 with that calculated in Step 8. What effect did adding R_3 to the circuit have on this value?

5. What do the results in Tables 7.4 and 7.6 tell you about the relationship between P_T and individual power values in a parallel circuit?

SIMULATION EXERCISE

Procedure

1. Open file Ex7.1 from the Electronics Technology Fundamentals companion web site (www.prenhall.com/paynter). Before running the simulation, check to make sure that keyboard control of the switches has been established.
2. Start the simulation with all three switches open. Does the I_T meter indicate a small amount of current? If so, what is the source of this current?

3. Record the current meter readings for each switch combination listed in Table 7.8. Leave the simulation running the entire time.

TABLE 7.8   Meter Readings

	Meter Readings			
Closed Switch(es)	I_T	I_1	I_2	I_3
S1 only				
S2 only				
S3 only				
S1 and S2				
S1, S2, and S3				

Faults

4. Stop the simulation, and open all three switches. Fault R_1 shorted.
5. Close S1, and describe what happens.

6. Remove the fault from R_1, and fault R_2 open.

7. With all switches open, start the simulation again. Close the switches in the order indicated in Table 7.9, and record the meter readings in the table.

TABLE 7.9

	Meter Readings			
Closed Switch(es)	I_T	I_1	I_2	I_3
S3 only				
S3 and S2				
S3, S2, and S1				

Questions

1. The ideal resistance of an open switch is _____. The ideal resistance of a closed switch is _____.

2. Refer to the first row of Table 7.8. When only one resistor is switched into the circuit, there is a slight difference between the I_T and I_1 meter readings. What is the source of the difference?

3. Do the meter readings in Table 7.8 support Kirchhoff's current law? Explain your answer by citing specific examples.

4. Based on your observations in Step 5, describe the effect of a shorted branch on parallel circuit operation.

5. Based on your measured values in Table 7.9, describe the effect of an open branch on parallel circuit operation.

6. Based on your observations, comment on the following statement: *When troubleshooting a parallel circuit, the condition of the fuse is generally a good indicator for the type of fault that is present.*

Exercise 8

Kirchhoff's Current Law and the Current-Divider Relationship

OBJECTIVES

After completing this exercise, you should be able to:

- Demonstrate Kirchhoff's current law.
- Apply Kirchhoff's current law to the current analysis of a parallel circuit.
- Calculate the value of an unknown resistor in a parallel circuit using Kirchhoff's current law.
- Demonstrate the concept of current division in parallel circuits.
- Demonstrate the concept of inverse proportionality as it applies to current division in parallel circuits.

DISCUSSION

In this exercise, you will examine Kirchhoff's current law and the current-divider relationship. As you may recall, Kirchhoff's voltage law is used to describe the relationship between the voltages in a series circuit. In the same manner, Kirchhoff's current law is used to describe the relationship between the currents in a parallel circuit.

A *node* is a junction. Kirchhoff's current law states that the sum of the currents entering a node must equal the sum of the currents leaving that node. You will investigate this relationship in this exercise. You will also investigate the current-divider relationship. In a series circuit, the voltage is divided among the various resistors in proportion to their relative values. The greater the value of a given resistor, the greater the voltage across its terminals. In a parallel circuit, current is divided among the various branches. Unlike voltage division, the division of current varies *inversely* with the branch resistance: the greater the resistance of a branch, the lower the portion of total current through that branch. This is another case in which the characteristics of parallel circuits appear opposite to those of series circuits.

LAB PREPARATION

Review Sections 5.3 through 5.5 of *Electronics Technology Fundamentals*.

MATERIALS

2   DMMs
1   variable dc power supply
1   protoboard
4   resistors: 2.2 kΩ, 3.3 kΩ, 4.7 kΩ, and 10 kΩ
1   10 kΩ potentiometer

PROCEDURE

1. Measure the value of each resistor, and enter its value in Table 8.1.

TABLE 8.1

Nominal Value	2.2 kΩ	3.3 kΩ	4.7 kΩ	10 kΩ
Measured Value				

2. Construct the circuit shown in Figure 8.1. Apply power, and measure the total circuit current. Record the value of I_T in Table 8.2.

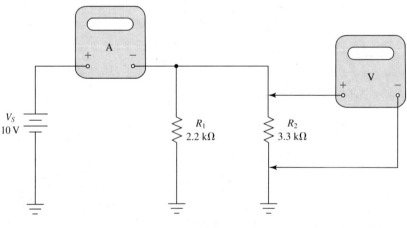

FIGURE 8.1

TABLE 8.2

I_T	V_{R1}	$\dfrac{V_{R1}}{R_1}$	V_{R2}	$\dfrac{V_{R2}}{R_2}$

3. Measure the voltage across each branch, and use that value to calculate each branch current. Enter your measured and calculated values in Table 8.2.

4. Calculate the sum of the branch currents that you recorded in Table 8.2.

$$I_{R1} + I_{R2} = \underline{\hspace{3cm}}$$

5. Add R_3 to the circuit as shown in Figure 8.2. Repeat Steps 2 and 3 for the new circuit. Record your measurements and calculations in Table 8.3.

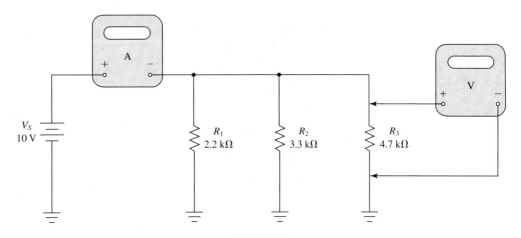

FIGURE 8.2

TABLE 8.3

I_T	V_{R1}	$\dfrac{V_{R1}}{R_1}$	V_{R2}	$\dfrac{V_{R2}}{R_2}$	V_{R3}	$\dfrac{V_{R3}}{R_3}$

6. Calculate the sum of the branch currents that you recorded in Table 8.3.

$$I_{R1} + I_{R2} + I_{R3} = \underline{\hspace{3cm}}$$

7. Take the 10 kΩ potentiometer, and rotate the control shaft to approximately halfway between its extremes. *Do not measure its value.* Replace the 4.7 kΩ resistor with the potentiometer as shown in Figure 8.3.

8. Apply power to the circuit, and measure I_T. Use Kirchhoff's current law to determine the value of I_{R3}. (Since V_S has not changed, you can assume that the values of I_{R1} and I_{R2} are the same as those listed in Table 8.3.)

$$I_{R3} = \underline{\hspace{3cm}}$$

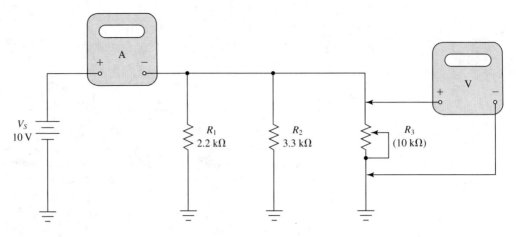

FIGURE 8.3

9. Use Ohm's law to calculate R_3.

$$R_3 = \underline{\hspace{2cm}} \text{ (calculated)}$$

10. Remove the pot, taking care not to change its setting, and then measure its value.

$$R_3 = \underline{\hspace{2cm}} \text{ (measured)}$$

11. Construct the circuit shown in Figure 8.4. Adjust the potentiometer until you get a current reading of 1 mA. (*Note:* The 10 kΩ potentiometer is being used to establish a specific value of total circuit current and is considered to be part of the source. Other than adjusting its value when directed, you should ignore this component.)

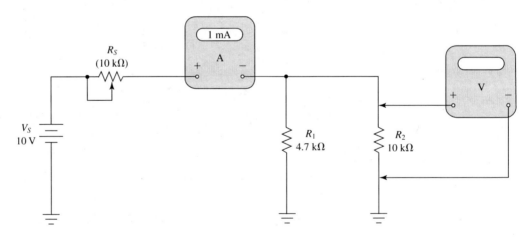

FIGURE 8.4

12. Measure the voltage across each branch. Use the measured voltages and Ohm's law to calculate the current in each branch, and enter your results in Table 8.4.

TABLE 8.4

I_T (measured)	V_{R1}	$\dfrac{V_{R1}}{R_1}$	V_{R2}	$\dfrac{V_{R2}}{R_2}$

13. Calculate the total resistance of the parallel circuit. (*Remember:* Ignore the presence of the potentiometer.) Then, use the current-divider equation to calculate the branch currents.

$$R_T = \underline{\hspace{3cm}}$$

$$I_{R1} = I_T \frac{R_T}{R_1} = \underline{\hspace{3cm}}$$

$$I_{R2} = I_T \frac{R_T}{R_2} = \underline{\hspace{3cm}}$$

14. Add R_3 to the circuit as shown in Figure 8.5. Adjust the potentiometer to return I_T to 1 mA.

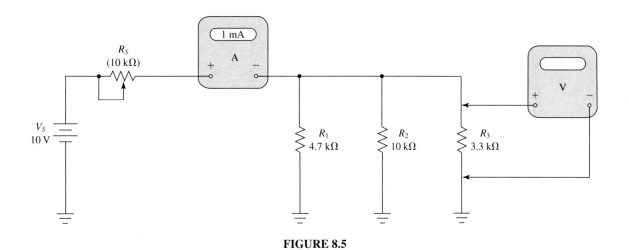

FIGURE 8.5

15. Measure the voltage across each branch. Use the measured voltages and Ohm's law to calculate the current in each branch, and enter your results in Table 8.5.

TABLE 8.5

V_{R1}	$\dfrac{V_{R1}}{R_1}$	V_{R2}	$\dfrac{V_{R2}}{R_2}$	V_{R3}	$\dfrac{V_{R3}}{R_3}$

16. Calculate the total resistance of the parallel circuit. (*Remember:* Ignore the presence of the potentiometer.) Then, use the current-divider equation to calculate the branch currents.

$$R_T = \underline{\hspace{3cm}}$$

$$I_{R1} = I_T \frac{R_T}{R_1} = \underline{\hspace{3cm}}$$

$$I_{R2} = I_T \frac{R_T}{R_2} = \underline{\hspace{3cm}}$$

$$I_{R3} = I_T \frac{R_T}{R_3} = \underline{\hspace{3cm}}$$

QUESTIONS & PROBLEMS

1. Did your results from Steps 2 and 4 conform to Kirchhoff's current law? Explain why or why not.

2. Refer to Figure 8.1 and Table 8.2. Calculate the ratio of R_1 to R_2. Now, calculate the ratio of I_{R2} to I_{R1}. Explain what, if anything, this tells you about current division in a parallel circuit.

$$\frac{R_1}{R_2} = \underline{\hspace{3cm}} \qquad \frac{I_{R2}}{I_{R1}} = \underline{\hspace{3cm}}$$

3. Refer to Steps 5 and 6. Did the results from these procedures conform to Kirchhoff's current law? Explain.

4. Refer to Steps 7 through 10. Explain how Kirchhoff's current law can be used to determine the value of an unknown resistor in a parallel circuit.

5. Refer to Steps 11 through 16. Perform the calculations shown below, then explain what these results tell you about the concept of inverse proportionality.

$$\frac{R_1}{R_T} = \underline{\hspace{2cm}} \qquad \frac{I_{R1}}{I_T} = \underline{\hspace{2cm}} \qquad \frac{I_T}{I_{R1}} = \underline{\hspace{2cm}}$$

$$\frac{R_2}{R_T} = \underline{\hspace{2cm}} \qquad \frac{I_{R2}}{I_T} = \underline{\hspace{2cm}} \qquad \frac{I_T}{I_{R2}} = \underline{\hspace{2cm}}$$

$$\frac{R_3}{R_T} = \underline{\hspace{2cm}} \qquad \frac{I_{R3}}{I_T} = \underline{\hspace{2cm}} \qquad \frac{I_T}{I_{R3}} = \underline{\hspace{2cm}}$$

SIMULATION EXERCISE

Discussion

There are several different types of sources. Though we most commonly deal with voltage sources, some electrical and electronic devices have the characteristics of a *current source*. In the hardware portion of this exercise, you were shown that a voltage source in series with a high-value resistance can effectively be substituted for a current source. This substitution is valid, but it does have limitations.

Compliance is defined as *the act of complying with a demand*. Voltage sources and current sources exhibit different forms of compliance. When the load connected to a voltage source changes, it results in a change in the source current. This is referred to as *current compliance*. When the load connected to a current source changes, it results in a change in the source voltage. This is referred to as *voltage compliance*.

In a previous simulation, you were shown that:

- A voltage source produces its rated output when its load is open.
- A voltage source must be protected from a shorted-load condition (with a fuse or current-limiting device).

In contrast:

- A current source produces its rated output when its load is shorted.
- A true current source must be protected from an open-load condition (with an alternate current path).

A true current source must be protected from an open load because it will attempt to produce a very high output voltage to comply with the open-load condition. As a result, *an open-circuited current source is inherently dangerous and is likely to be damaged as a result of the open*. To protect the current source (and the technician) from an open-load condition, an alternate current path must be provided to limit the source output voltage. (*Note:* A true current source is maintained with a short across the terminals until it is ready to be used, at which time the short is removed).

As you may have observed, voltage and current sources are *opposites* in terms of their response to changes in load. As a result, it is easy to become confused when working with a current source, which can lead to injury or equipment damage. Simulations allow you to investigate current source characteristics both safely and effectively.

Procedure

1. Open file Ex8.1 from the Electronics Technology Fundamentals companion web site (www.prenhall.com/paynter). Note that initially, all of the switches are closed.
2. Open and close the switches in the order listed in Table 8.6, then make the appropriate measurements to complete the table.

TABLE 8.6

Switch States		V_S	I_T	I_{R1}	I_{R2}	I_{R3}	I_{R4}
Open	Closed						
None	All						
S_1	S_2, S_3, S_4						
S_1, S_2	S_3, S_4						
S_2, S_3	S_1, S_4						
S_1, S_2, S_3	S_4						
All	None						

Questions

1. Citing examples from Table 8.6, show that in all cases (including the S_4 only closed condition), Kirchhoff's current law is validated.

2. Refer to the V_S readings in Table 8.6. Explain how these readings illustrate the concept of *voltage compliance,* as outlined in the discussion.

3. Refer to the V_S reading in Table 8.6 for all switches open. Comment on the following statement: "This reading shows that the current source in the simulator is not ideal and has limitations."

Exercise 9

Series-Parallel Circuits

OBJECTIVES

After completing this exercise, you should be able to:

- Analyze series circuits connected in parallel.
- Analyze parallel circuits connected in series.

DISCUSSION

In the real world, few circuits are purely series or purely parallel. Most circuits are some combination of the two. The simplest series-parallel circuits contain either of the following:

- Two or more series circuits connected in parallel.
- Two or more parallel circuits connected in series.

More complex circuits may contain any combination of series and parallel connections, which makes it difficult to determine whether components are connected in series or in parallel. This is why it is so important to understand the basics of series and parallel circuits. No matter how complex the circuit, it can be reduced to a manageable *equivalent circuit* if you take your time and remember your basics.

LAB PREPARATION

Review Sections 6.1 and 6.2 of *Electronics Technology Fundamentals*.

MATERIALS

2   DMMs
1   variable dc power supply
1   protoboard
5   resistors: 220 Ω, 330 Ω, 470 Ω, 510 Ω, and 1 kΩ

PROCEDURE

> *Note:* In most of the exercises, you have measured the value of each resistor. From this point on, you will simply work with the nominal values in your calculations. If your calculated and measured values differ by a small amount, say ±5%, then assume that component tolerance is the culprit. If the disparity is greater, however, you should then recheck your measurements and calculations.

1. Refer to the circuit shown in Figure 9.1. Calculate the equivalent resistance for each parallel circuit, and use these values to calculate total circuit resistance. Then, solve for I_T and record its value in Table 9.1.

$R_A = $ _____ $R_B = $ _____ $R_T = $ _____

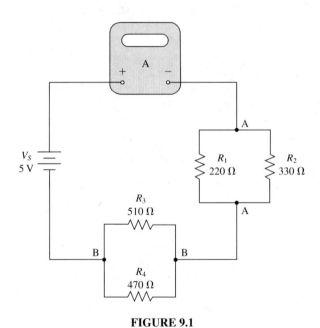

FIGURE 9.1

TABLE 9.1

Value	Calculated	Measured	Variation (%)
I_T			
V_A			
V_B			

2. Using your calculated values of R_A and R_B and the voltage-divider rule, calculate the voltage across each parallel network. Enter your results in Table 9.1.
3. Construct the circuit shown in Figure 9.1. Measure the total circuit current and the voltage across each branch. Enter all your measurements in Table 9.1.
4. Calculate the percent of variation between your measured and calculated values, and enter your results in Table 9.1.
5. Refer to the circuit shown in Figure 9.2. Calculate the equivalent resistance for each branch and the total circuit resistance.

$R_A =$ _____ $R_B =$ _____ $R_T =$ _____

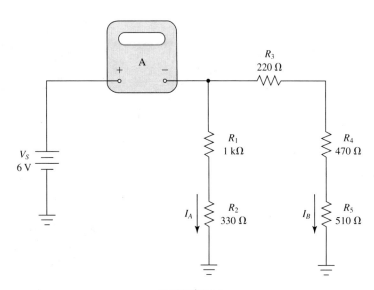

FIGURE 9.2

6. Based on your results in Step 5, calculate the circuit currents and component voltages. Enter your calculated values in Table 9.2.

TABLE 9.2

Value	Calculated	Measured	Variation (%)
I_T			
I_A			
I_B			
V_{R1}			
V_{R2}			
V_{R3}			
V_{R4}			
V_{R5}			

7. Construct the circuit shown in Figure 9.2. Measure the total circuit current and component voltages. Use your measured voltages and nominal resistor values to calculate the values of the branch currents, and enter your results in Table 9.2.

8. Calculate the percent of variation between your calculated and measured values, and enter your results in Table 9.2.

QUESTIONS & PROBLEMS

1. Did your results from Steps 1 through 4 conform with what you have learned about parallel circuits connected in series? Explain your answer.

2. Refer to your results in Table 9.1. Were any differences between your measured and calculated values outside the range of your component tolerances? Comment on why or why not.

3. Did your results from Steps 5 through 8 conform with what you have learned about series circuits connected in parallel? Explain your answer.

4. Refer to your results in Table 9.2. Did any differences between your measured and calculated values exceed the range of your component tolerances? Comment on why or why not.

5. Refer to the circuit shown in Figure 9.2. Assume that a resistor is added in parallel across R_4 and R_5. Would R_3 dissipate more or less power? Would I_{R5} increase or decrease? Explain your answers.

SIMULATION EXERCISE

Procedure

1. Open file Ex9.1 from the Electronics Technology Fundamentals companion web site (www.prenhall.com/paynter). This is the circuit shown in Figure 9.1, but with more instrumentation.
2. Run the simulation and complete the first row of Table 9.3. Compare these results to those in Table 9.1 to confirm that the circuit is operating normally.

TABLE 9.3

Condition	I_T	V_A	V_B
Normal			
R_2 out of tolerance			
R_2 shorted			
R_4 open			

Fault Symptoms

3. Simulate a partial failure of R_2 by changing its value to 33 Ω.
4. Run the simulation again with the faulted R_2. Record the ammeter and voltmeter readings in Table 9.3.
5. Fault R_2 again, but this time to a short.
6. Run the simulation again with the faulted R_2. Record the ammeter and voltmeter readings in Table 9.3.

7. Return R_2 to its original unfaulted value, and fault R_4 to the open condition.
8. Run the simulation again with the faulted R_4. Record the ammeter and voltmeter readings in Table 9.3.
9. Open file Ex9.2 from the Electronics Technology Fundamentals companion web site (www.prenhall.com/paynter). This is the circuit shown in Figure 9.2, but with quite a bit more instrumentation. Run the simulation and complete the first row of Table 9.4. Compare these values to those in Table 9.2 to confirm that the circuit is operating normally.

TABLE 9.4

Conditions	I_T	I_A	I_B	V_{R1}	V_{R2}	V_{R3}	V_{R4}	V_{R5}
Normal								
R_1 open								
R_5 shorted								

10. Stop the simulation and fault R_1 open. Run the simulation again and complete the second row of Table 9.4.
11. Stop the simulation and return R_1 to its normal condition. Then fault R_5 as a short. Run the simulation again and complete the third row of Table 9.4.
12. Compare the voltage and current indications of the faulted circuits with those of the normal circuit, and record your observations below.

Questions

1. Comment on the similarities and differences between the normal voltage and current readings and those for the partial failure of R_2 in Table 9.3. You should consider how all the circuit values were changed by the fault.

2. Comment on the similarities and differences between the normal voltage and current readings and those for R_2 shorted in Table 9.3. Consider how all the circuit values were changed by the fault.

3. Comment on the similarities and differences between the normal voltage and current readings and those for R_4 open in Table 9.3. Consider how all the circuit values were changed by the fault.

4. For a circuit like the one shown in Figure 9.1, does a problem in one of the parallel circuits affect the other parallel combination? How, and why?

5. In Steps 10 through 12, a circuit like the one shown in Figure 9.2 was faulted.
   a. Describe the effect of an open in one branch on the values in the other branch.

   b. Describe the effect of an open in one branch on the values in that branch.

c. Describe the effect of a short in one branch on the values in the other branch.

d. Describe the effect of a short in one branch on the values in the branch containing the short.

6. Does Kirchhoff's current law hold true for circuits with faults? Cite specific examples from Tables 9.3 and 9.4 in your answer.

7. Does Kirchhoff's voltage law hold true for circuits with faults? Cite specific examples from Tables 9.3 and 9.4 in your answer.

Exercise 10

Loaded Voltage Dividers

OBJECTIVES

After completing this exercise, you should be able to:

- Demonstrate the concept of circuit loading.
- Demonstrate the concept of voltage-divider stability.
- Predict the effect that a change in load demand has on load voltage.
- Explain the concept of no-load and full-load conditions.

DISCUSSION

In this exercise, you will examine the effects of loading on the operation of a voltage divider. A loaded voltage divider (like the one shown in Figure 10.1) is commonly used to establish a specific voltage that is lower in magnitude than the supply voltage. The output from a loaded voltage divider (V_L) is always less than the circuit's *no-load output voltage* (V_{NL}). In this exercise, you will take a closer look at this relationship.

It is not uncommon for the value of a load to vary—sometimes significantly. For example, some speakers are 8 Ω, some are 4 Ω, and some can dip as low as 2 Ω. When a speaker is connected to the output of an amplifier, the amplifier output acts, in effect, as a loaded voltage divider. As a result, the resistance of the speaker can significantly affect the operation of the amplifier.

The *stability* of a voltage divider is determined by how much its output voltage changes when the value of its load varies. As you will observe in this exercise, the relationship between voltage-divider *bleeder current* and load current has a significant effect on the circuit's stability.

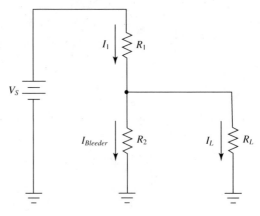

FIGURE 10.1

LAB PREPARATION

Review Section 6.3 of *Electronics Technology Fundamentals*.

MATERIALS

1   DMM
1   variable dc power supply
1   protoboard
9   resistors: 100 kΩ, 47 kΩ, 10 kΩ (2), 4.7 kΩ, 3.3 kΩ, 1 kΩ (2), and 100 Ω
1   10 kΩ potentiometer

PROCEDURE

1.   Calculate the no-load output voltage and bleeder current for the circuit shown in Figure 10.2. Enter your calculations in Table 10.1.

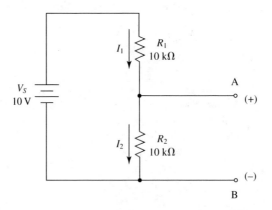

FIGURE 10.2

TABLE 10.1   Open-Load Values for the Circuit in Figure 10.2

Value	Calculated	Measured
V_{NL}		
I_2		

2. Construct the circuit shown in Figure 10.2. Measure the no-load output voltage, and use the measured value of V_{NL} to calculate the bleeder current (I_2). Enter your results in Table 10.1.
3. Connect a 100 kΩ load to the circuit. Apply power, and measure the load voltage (V_L). Use V_L to calculate:
   • The load and bleeder currents.
   • The percent change in V_L from its no-load value.
   Enter your results in Table 10.2.
4. Repeat Step 3 for loads of 47 kΩ, 1 kΩ, and 100 Ω. Enter your results in Table 10.2.

TABLE 10.2   Values for the Circuit in Figure 10.2

R_L	V_L	% Change	I_L	I_2
100 kΩ				
47 kΩ				
1 kΩ				
100 Ω				

5. Replace R_1 and R_2 with two 1 kΩ resistors. (*Note:* This should not change the value of V_{NL}, because the ratio of R_1 to R_2 has not changed.) Repeat Steps 3 and 4 for this new circuit, and enter your results in Table 10.3.

TABLE 10.3   Values for the Modified Figure 10.2 Circuit ($R_1 = R_2 = 1$ kΩ)

R_L	V_L	% Change	I_L	I_2
100 kΩ				
47 kΩ				
1 kΩ				
100 Ω				

6. Figure 10.3 shows a potentiometer being used as a variable voltage divider. Construct this circuit, but do not connect the load. Initially, the resistance from the wiper arm to ground should be set to its maximum value. Apply power to the circuit, and adjust the potentiometer so that the voltage at the wiper arm (V_{BC}) is 3 V. This is the no-load output voltage (V_{NL}) for the circuit.

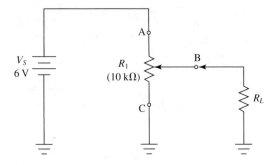

FIGURE 10.3

7. Remove power, and connect a 10 kΩ load to the wiper arm. Apply power, and adjust the potentiometer until $V_L = 2.7$ V, which is 90% of V_{NL}. (*Note:* A voltage divider is considered to be stable if its output varies by no more than 10% from its no-load voltage when a load is connected.)

8. Remove power, and disconnect the potentiometer from the circuit without changing the wiper arm setting. Then:

   • Measure R_{BC}.

   • Use the adjusted load voltage to calculate the load and bleeder currents.

   • Calculate the ratio of bleeder current to load current.

   Enter your results in Table 10.4.

TABLE 10.4

R_L	R_{BC}	I_L	$I_{Bleeder}$	$\dfrac{I_{Bleeder}}{I_L}$
10 kΩ				
4.7 kΩ				
3.3 kΩ				

9. Repeat Steps 6 through 8 for loads of 4.7 kΩ and 3.3 kΩ, and enter your results in Table 10.4.

QUESTIONS & PROBLEMS

1. Refer to Table 10.2. Why did the percent of change in output voltage increase as the load resistance decreased?

2. Compare your results in Tables 10.2 and 10.3. Which of the two circuits was more stable? Why do you think this was the case?

3. Calculate the no-load power dissipation for the circuits used to complete Tables 10.2 and 10.3. Which circuit dissipates the most power when the load is open? Keeping in mind your power values and your answer to Question 2 above, comment on the relationship between circuit efficiency and circuit stability.

4. Refer to your results in Table 10.4. What do your calculated current ratios tell you about voltage-divider stability?

SIMULATION EXERCISE

Procedure

1. Open file Ex10.1 from the Electronics Technology Fundamentals companion web site (www.prenhall.com/paynter). This is the circuit shown in Figure 10.2, but with a 10 MΩ load attached (which is very close to a no-load condition), and more instrumentation. Note that all the voltage indicators have internal resistance values of 10 MΩ.

2. Run the simulation, and record the meter readings in Table 10.5.

TABLE 10.5

Condition	I_T	I_2	I_L	V_{R1}	V_L
$R_L = 10 \text{ M}\Omega$					
$R_L = 10 \text{ k}\Omega$					
R_L open					
R_L shorted					
R_2 open					
R_2 shorted					
R_1 open					
R_1 shorted					

3. Change the value of the load resistor to 10 kΩ. Run the simulation, and record the meter readings in Table 10.5.
4. Fault the load resistor to the open condition. Run the simulation, and record the meter readings in Table 10.5.
5. Fault the load resistor to the shorted condition. Run the simulation, and record the meter readings in Table 10.5.
6. Restore the load resistance to 10 kΩ, and fault the bleeder resistor (R_2) to the open condition. Run the simulation, and record the meter readings in Table 10.5.
7. Fault the bleeder resistor (R_2) to the shorted condition. Run the simulation, and record the meter readings in Table 10.5.
8. Restore R_2 to its normal value, and fault R_1 to the open condition. Run the simulation, and record the meter readings in Table 10.5.
9. Fault R_1 to the shorted condition. Run the simulation, and record the meter readings in Table 10.5.

Questions

1. Compare the readings that were obtained with an open load to those resulting from a load resistor value of 10 MΩ. How much difference is there between the readings?

2. Based on your answer to Question 1, what conclusion can you draw about the significance of the internal resistance of the voltage indicators?

3. Why did both the shorted load and the shorted bleeder resistor produce a 0 V output? What was the difference in current indications?

4. With R_1 open, the simulation shows slight currents through all the current indicators. Explain the presence of this current, and identify the current paths. Under what condition would this current decrease to 0 A?

5. *In the loaded voltage divider, bleeder voltage and load voltage are always equal.* This statement is an application of which principle?

6. Explain the readings that were produced when R_1 was shorted. Comment on any additional problems that might occur in a circuit if this resistor is shorted.

Exercise 11

The Superposition Theorem

OBJECTIVES

After completing this exercise, you should be able to:

- Analyze a two-source circuit using the superposition theorem.
- Write the Kirchhoff loop equations for a two-source circuit.

DISCUSSION

So far, all the circuits you have analyzed have been single-source circuits. Some circuits, which are referred to as *multisource circuits,* have two or more voltage (and/or current) sources. The circuit shown in Figure 11.1 is an example of a multisource circuit. The *superposition theorem*

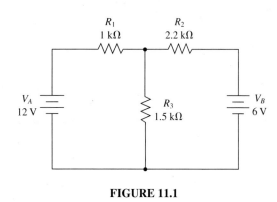

FIGURE 11.1

states that the response of a circuit to more than one source can be determined by analyzing the circuit's response to each source alone and then combining the results.

According to the superposition theorem, the component voltages in a circuit containing two voltage sources can be determined as follows:

1. Replace one of the voltage sources with an equivalent resistance.
2. Determine the magnitude and polarity of the voltage across each circuit component.
3. Repeat Steps 1 and 2, this time restoring the first voltage source and replacing the second voltage source with an equivalent resistance.
4. Add all the component voltages, taking their respective polarities into account.

In this exercise, you will use the superposition theorem to predict the component voltages for the circuit shown in Figure 11.1. You will also demonstrate that Kirchhoff's voltage law is applicable to any closed loop.

LAB PREPARATION

Review Section 7.1 of *Electronics Technology Fundamentals*.

MATERIALS

1   DMM
1   dual-output variable dc power supply
1   protoboard
4   resistors: 2.2 kΩ, 1.5 kΩ, and 1 kΩ (2)

PROCEDURE

1. Construct the circuit shown in Figure 11.2. Note that V_B in Figure 11.1 has been replaced with a short circuit. This short circuit is used to simulate the ideal internal resistance of that voltage source.

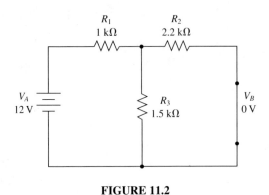

FIGURE 11.2

2. Measure the voltage across each resistor, and record these measurements in Table 11.1. Mark the polarity of each resistor voltage on the components in Figure 11.2.

TABLE 11.1

Condition	V_{R1}	V_{R2}	V_{R3}
V_A only			
V_B only			
$V_A + V_B$ (calculated)			
$V_A + V_B$ (measured)			
Variation (%)			

3. Disconnect power from the circuit. Reconnect the voltage source (V_B), and replace V_A with a short, as shown in Figure 11.3.

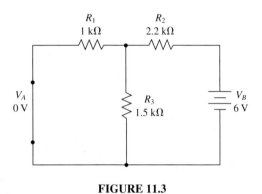

FIGURE 11.3

4. Repeat Step 2 for the new circuit, but mark the polarity of each resistor voltage on Figure 11.3.
5. Based on your results from Steps 3 and 4, determine the sum of the individual resistor voltages for the two circuits. Don't forget to take the relative polarities of the component voltages into account. Enter your calculations in Table 11.1.
6. Construct the original multisource circuit shown in Figure 11.1. Measure the resistor voltages, and record your measurements in Table 11.1.
7. Determine the percent of variation between your calculated and measured values.
8. Refer to the circuit shown in Figure 11.4. This is the same circuit as shown in Figure 11.1, but the polarities of the voltage sources have been reversed.

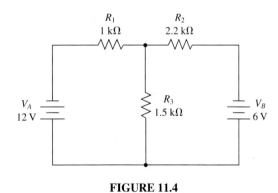

FIGURE 11.4

9. Repeat Steps 1 through 7 for the circuit shown in Figure 11.4, recording your measured and calculated values in Table 11.2.

TABLE 11.2

Condition	V_{R1}	V_{R2}	V_{R3}
V_A only			
V_B only			
$V_A + V_B$ (calculated)			
$V_A + V_B$ (measured)			
Variation (%)			

QUESTIONS & PROBLEMS

1. Refer to your results in Table 11.1. Did these results substantiate the superposition theorem? Explain your answer.

2. Refer to your results in Table 11.2. Did these results substantiate the superposition theorem? Explain your answer.

3. Refer to the circuit shown in Figure 11.1 and your results in Table 11.1. Using your measured results, demonstrate that each loop conforms to Kirchhoff's voltage law.

4. Refer to the circuit shown in Figure 11.4 and your results in Table 11.2. Using your measured results, demonstrate that each loop conforms to Kirchhoff's voltage law.

SIMULATION EXERCISE

Procedure

1. Open file Ex11.1 from the Electronics Technology Fundamentals companion web site (www.prenhall.com/paynter). This is the circuit shown in Figure 11.1, but with more instrumentation. We will refer to this circuit as Circuit 1. Note that all the voltage indicators have internal resistance values of 10 MΩ.
2. Run the simulation and complete row 4 of Table 11.3.

TABLE 11.3

Condition	V_{R1}	V_{R2}	V_{R3}	I_{R1}	I_{R2}	I_{R3}
V_A only (Circuit 2)						
V_B only (Circuit 3)						
$V_A + V_B$ (calculated)						
$V_A + V_B$ (Circuit 1)						
Variation (%)						

3. Now, change V_B to 0 V. This circuit is now the equivalent of Figure 11.2. We will refer to this circuit as Circuit 2. It should be noted that the voltage sources in the simulator are ideal, so their internal resistance is 0 Ω.

4. Run the simulation and complete row 1 of Table 11.3.

5. Restore V_B to 6 V and change V_A to 0 V. This circuit is now the equivalent of Figure 11.3. We will refer to this circuit as Circuit 3.

6. Run the simulation and complete row 2 of Table 11.3.

7. Add the resistor voltage values from Circuits 2 and 3. Remember to take voltage polarity into account. Record your results in the third row of Table 11.3. Then, calculate the percent of variation between your calculated values in row 3 and the measured values in row 4 of the table. Record your results in the bottom row of the table.

Questions

1. The sources in a simulator are ideal sources with an internal resistance of zero. Setting the source value to zero effectively removes the source. What other approach could be used to achieve the same result?

2. If you were trying to run a circuit simulation and knew the value of the source resistance, how would you amend Circuits 2 and 3? Under what circumstance could the actual source resistance be ignored?

3. Using Ohm's law and Kirchhoff's current law, explain the ammeter readings in Table 11.3.

Exercise 12

Thevenin's Theorem and Maximum Power Transfer

OBJECTIVES

After completing this exercise, you should be able to:

- Derive the Thevenin equivalent of a series-parallel circuit.
- Determine load voltages for changing load values using Thevenin equivalent circuits.
- Demonstrate the maximum power transfer theorem using Thevenin equivalent circuits.

DISCUSSION

One common concern in electronics is predicting the response of a given circuit to a change in load. Using conventional techniques, you would need to do a complete circuit analysis for each new load value to make such a prediction. Thevenin's theorem, however, allows you to represent any resistive circuit as an equivalent circuit that contains a single voltage source in series with a single resistance. Using this *Thevenin equivalent circuit,* predicting the changes in load voltage that result from changes in load resistance becomes nothing more than a series of simple voltage-divider calculations.

Thevenin equivalent circuits also allow you to easily determine the load resistance that will allow maximum power transfer from the source to that load. The maximum power transfer theorem states that when the source resistance is fixed and the load is variable, maximum power transfer occurs when the load resistance equals the source resistance. This means that maximum power is transferred from a source to its load when $R_L = R_{TH}$. In this exercise, you will examine this theorem.

LAB PREPARATION

Review Sections 7.3 and 7.4 of *Electronics Technology Fundamentals*.

MATERIALS

1   DMM
1   decade resistance box
1   dual-output variable dc power supply
1   protoboard
7   resistors: 4.7 kΩ, 2.2 kΩ, 2 kΩ, and 1 kΩ (4)

PROCEDURE

1. Calculate the Thevenin equivalent values for the test circuit shown in Figure 12.1(a).

$$V_{TH} = \rule{3cm}{0.4pt}$$

$$R_{TH} = \rule{3cm}{0.4pt}$$

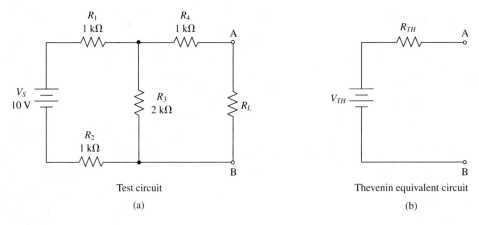

Test circuit

(a)

Thevenin equivalent circuit

(b)

FIGURE 12.1

2. Construct the Thevenin equivalent of the test circuit. Set the dc power supply to the value of V_{TH} that you calculated in Step 1. Use the decade box to produce the value of R_{TH} that you calculated in the same step.
3. Connect a 1 kΩ load to the Thevenin equivalent circuit. Measure the load voltage, and enter your measurement in Table 12.1.

TABLE 12.1   Measured Load Voltages for the Circuits Shown in Figure 12.1

Load	V_L		Variation (%)
	Thevenin Circuit	Test Circuit	
$R_L = 1\ \text{k}\Omega$			
$R_L = 2.2\ \text{k}\Omega$			
$R_L = 4.7\ \text{k}\Omega$			

4. Repeat Step 3 for load values of 2.2 kΩ and 4.7 kΩ.
5. Construct the test circuit shown in Figure 12.1. Repeat Steps 3 and 4, and enter your results in Table 12.1.
6. Using the test circuit values as a reference, calculate the percent of variation in the load voltages that you obtained from the Thevenin equivalent circuit.
7. According to the maximum power transfer theorem, maximum power transfer from a source to a variable load occurs when $R_L = R_{TH}$. Using the decade box as a load, set R_L for the test circuit to the value of R_{TH} that you calculated in Step 1. Measure the voltage across the load, and enter the values of R_L and V_L in the appropriate spaces of Table 12.2.

TABLE 12.2   Test Circuit Load Voltage and Power at Specified Values of R_L

Load Values	R_L	V_L	P_L
$R_{TH} + 50\%$			
$R_{TH} + 25\%$			
$R_{TH} + 10\%$			
R_{TH}			
$R_{TH} - 10\ \%$			
$R_{TH} - 25\%$			
$R_{TH} - 50\%$			

8. Repeat Step 7 for all the load resistance values listed in Table 12.2. Once again, use the decade box to produce the load values indicated, and enter your values for R_L and V_L in the table.
9. Calculate the power dissipated by the load for each value of R_L in Table 12.2, and enter your results in the table.

10. Using your results in Table 12.2, plot a curve of P_L versus R_L in Figure 12.2.

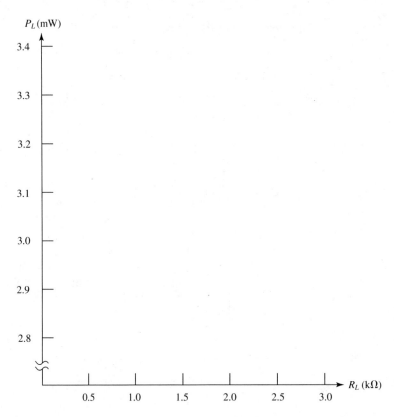

FIGURE 12.2   Power versus load resistance for the test circuit shown in Figure 12.1.

QUESTIONS & PROBLEMS

1. Refer to Table 12.1. Did your results substantiate Thevenin's theorem? Explain your answer. Also, explain any variations between your results for the test circuit and its Thevenin equivalent.

2. Refer to your results from Table 12.2. Did these results substantiate the maximum power transfer theorem? Explain your answer.

3. Refer to the circuit shown in Figure 12.1, and assume that R_4 has doubled in value. Explain why the Thevenin equivalent voltage for this circuit would not change.

SIMULATION EXERCISE

Procedure

1. Open file Ex12.1 from the Electronics Technology Fundamentals companion web site (www.prenhall.com/paynter). There are four circuits on your workspace. Circuit 4 represents the complete circuit with the load connected. As you can see, the circuit used in this simulation is slightly more complex than the circuit used in the hardware portion of this exercise.
2. Run the simulation and use Circuit 1 to measure R_{TH} and use Circuit 2 to measure V_{TH}. Record these values below and then stop the simulation.

$$R_{TH} = \text{_____} \qquad V_{TH} = \text{_____}$$

3. Change the values of R_{TH} and V_{TH} in Circuit 3 to those measured in Step 2.
4. Run the simulation and compare the load voltages between Circuits 3 and 4. Record these values below.

$$V_{L(\text{Circuit 3})} = \text{_____} \qquad V_{L(\text{Circuit 4})} = \text{_____}$$

5. Open file Ex12.2 from the Electronics Technology Fundamentals companion web site (www.prenhall.com/paynter). You will use this circuit to substantiate the maximum power transfer theorem. A piece of test equipment will be used in this simulation that you have not used before. It is a **wattmeter,** which is used to measure power. Note that the wattmeter is connected both in *series* with (to measure current), and in *parallel* across (to measure voltage) the load. This is consistent with the power equation $P = V \times I$.
6. As you can see, the load is an 8.4 kΩ variable resistor set to 50% of its maximum value. This should be very close to the value of R_{TH} that you measured in Step 2. Run the simulation and measure the load power. Record this value in the appropriate place in Table 12.3.

TABLE 12.3 Load Power for Changing Load Values

Load Resistance	Load Power
35%	
40%	
45%	
50% (R_{TH})	
55%	
60%	
65%	

7. Now, adjust the load to the values listed in Table 12.3 and complete the table.

Questions

1. Refer to Circuit 4 in File Ex12.1. Calculate the Thevenin values for this circuit and compare them to your measured values from Step 2. How do these values compare? Comment on any discrepancies.

2. Refer to your results from Step 4. Did the two circuits have the same load voltage? Explain how this supports (or refutes) Thevenin's theorem.

3. Refer to your results from Table 12.3. Are these results consistent with the maximum power transfer theorem? Explain your answer.

Part 2

AC Principles

Exercise 13

Inductors and Inductive Reactance (X_L)

After completing this exercise, you should be able to:

- Explain the relationship between voltage and current in an inductive circuit.
- Explain the relationship between inductive reactance and frequency.
- Determine the inductive reactance of a coil using Ohm's law.

DISCUSSION

In a purely resistive ac circuit, current and voltage are always in phase. In a purely inductive circuit, voltage leads current by 90°, as illustrated in Figure 13.1.

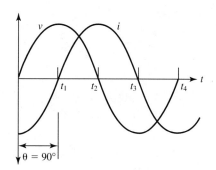

FIGURE 13.1 Phase relationship between inductor voltage and current.

To measure this phase relationship, you need a way to *effectively* measure current with an oscilloscope (which cannot display current waveforms). Here is our solution to the problem: Note the *sensing resistor* (labeled R_S) in Figure 13.2. The sensing resistor allows you to measure a low-amplitude voltage with the oscilloscope that is in phase with circuit current. (*Remember:* Resistor current and voltage are always in phase.) The value of R_S is chosen so that $R_S \ll X_L$. Fulfilling this relationship ensures that the circuit phase characteristics are very close to those of a purely inductive circuit.

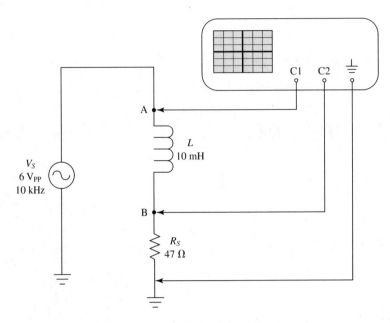

FIGURE 13.2 Inductive test circuit.

As you know, reactance changes when frequency changes. In the second part of this exercise, you will examine the relationship between inductive reactance and frequency.

LAB PREPARATION

Review:

- Sections 10.1, 10.2, and 10.4 of *Electronics Technology Fundamentals*.
- Your operator's manual on the proper use of the function generator.
- Appendix C on the proper use of the dual-trace oscilloscope.

MATERIALS

1	dual-trace oscilloscope
1	function generator
1	protoboard
2	resistors: 1 kΩ and 47 Ω
1	10 mH inductor

PROCEDURE

Part 1: Current and Voltage Phase Relationships

1. Construct the circuit shown in Figure 13.2. Connect Channel 1 (C1) of the scope to point A. This will allow you to monitor the voltage across the entire circuit. Since $R_S << X_L$ at the operating frequency of the circuit, the voltage at point A (with respect to ground) can be assumed to be approximately equal to V_L.

2. Connect Channel 2 (C2) of the scope to point B. This will allow you to measure the voltage across the sensing resistor. We know that V_{RS} is in phase with the circuit current. Since R_S is in series with the inductor, this voltage measurement has the same phase as the inductor current. By simultaneously displaying both waveforms, we can observe the phase relationship between V_L (Channel 1) and I_L (Channel 2).

3. Set the output of the signal generator for a 6 V_{PP} sine wave at a frequency of 10 kHz. Trigger on Channel 1. Next, adjust the vertical sensitivity (*volts/div*) of Channel 2 so that the two waveform displays are approximately equal in size. (They should appear similar to those shown in Figure 13.1.)

4. Draw the two waveforms to scale on the grid in Figure 13.3.

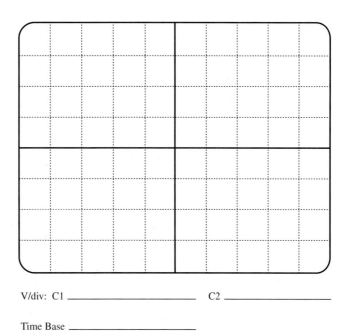

V/div: C1 _____   C2 _____

Time Base _____

FIGURE 13.3 Waveforms for the circuit shown in Figure 13.2.

5. The phase angle between two waveforms is measured as shown in Figure 13.4. Using the values of *t* and the period of the waveform (*T*), the phase angle between the waveforms is found using

$$\theta = (360°)\frac{t}{T}$$

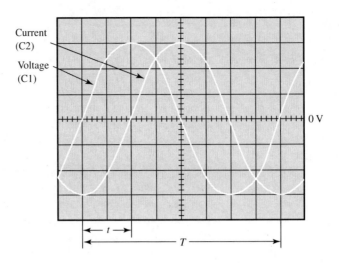

FIGURE 13.4 Phase angle measurement.

Using this equation, calculate the phase angle between the inductor voltage and current waveforms in Figure 13.3.

$$\theta = \underline{\hspace{2cm}}$$

Part 2: Inductive Reactance and Frequency

6. Construct the circuit shown in Figure 13.5. Set the output of the signal generator for an 8 V_{PP} sine wave at a frequency of 10 kHz. Use Channel 1 of the oscilloscope to monitor the output of the generator. Use Channel 2 to measure the voltage across R_S.

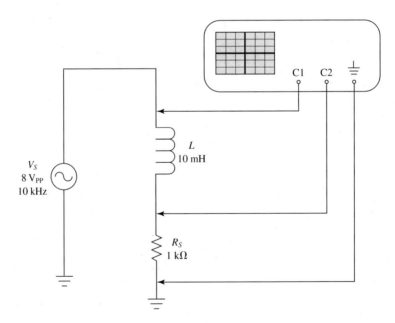

FIGURE 13.5

7. Measure the peak-to-peak value of V_{RS}, and record your measurement in Table 13.1.

TABLE 13.1   Measurements for the Circuit Shown in Figure 13.5

Frequency	V_{RS}	I_T	V_L	$X_L = \dfrac{V_L}{I_T}$	$X_L = 2\pi fL$
10 kHz					
20 kHz					
40 kHz					
60 kHz					
80 kHz					
100 kHz					

8. Repeat Step 7 for frequencies of 20 kHz, 40 kHz, 60 kHz, 80 kHz, and 100 kHz. Make certain that the signal applied to the circuit remains at 8 V_{PP} when you change frequencies. Enter your results in Table 13.1.

9. Reverse the position of the components as shown in Figure 13.6. Use Channel 1 to monitor the applied voltage and Channel 2 to measure the voltage across the inductor. Repeat Steps 7 and 8, and record the measured values of V_L in Table 13.1.

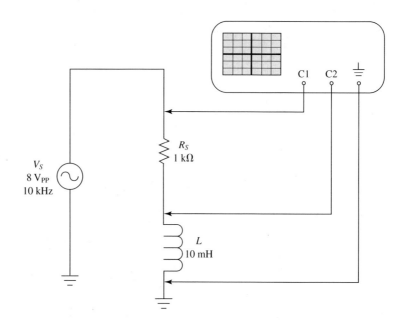

FIGURE 13.6

10. Using Ohm's law, calculate I_T for the circuit at each frequency, and enter your calculations in Table 13.1.

11. Using the values of I_T and V_L in Table 13.1, calculate the value of X_L at each frequency, and enter your results in Table 13.1.

12. Calculate X_L at each frequency using $X_L = 2\pi fL$. Enter your calculations in Table 13.1.

13. Using the values in the first and last columns in Table 13.1, plot a curve of inductive reactance versus frequency in Figure 13.7.

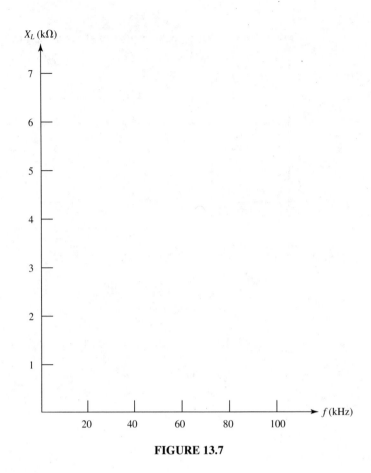

FIGURE 13.7

QUESTIONS & PROBLEMS

1. Refer to the two waveforms in Figure 13.3. Explain how they verify the current-versus-voltage phase relationship in an inductive circuit.

2. Explain how using Channel 1 as a trigger source (vs. using Channel 2) affects the oscilloscope display.

3. Refer to the two waveforms in Figure 13.3. Do the two waveforms appear to be *exactly* 90° out of phase? How do you explain any variation (from 90°) in the phase relationship?

4. Refer to your results in Table 13.1 and your graph in Figure 13.7. Explain what these results tell you about the relationship between frequency and inductive reactance.

5. Refer to your results in Table 13.1. Did your two methods of calculating X_L agree with each other? Explain any differences in the results.

6. Refer to your results in Table 13.1. Explain the relationship between the circuit current and operating frequency.

7. Refer to Figures 13.5 and 13.6. Explain why it was necessary to reverse the component positions to measure V_L.

8. According to the curve in Figure 13.7, did inductive reactance change with frequency at a linear or a nonlinear rate? Explain your answer.

SIMULATION EXERCISE

Procedure

1. Open file Ex13.1 from the Electronics Technology Fundamentals companion web site (www.prenhall.com/paynter). This is the circuit shown in Figure 13.2. In this simulation you will change component values and fault some components. Then you will see what effect these changes have on circuit operation.
2. Run the simulation, and record the following results in Table 13.2:
   a. The peak-to-peak value of V_{RS}.
   b. The phase angle between V_{RS} and V_S.
3. Fault the circuit by changing R_S to 470 Ω, and run the simulation again.
4. Measure and record the peak-to-peak resistor voltage and the phase angle between V_S and V_{RS}. Note any variations between these readings and those for normal circuit operation.
5. Restore the 47 Ω resistor, and fault the circuit by changing the inductor value to 1 mH. Run the simulation again.
6. Repeat Step 4.
7. Change the inductor value to 100 mH, and run the simulation again.
8. Repeat Step 4.
9. Fault the inductor open, and run the simulation again.
10. Repeat Step 4.
11. Restore the inductor to its original value, and fault the 47 Ω resistor open. Run the simulation again.
12. Repeat Step 4.

TABLE 13.2

Condition	V_{RS}	θ	Variation from Normal Operation
Normal operation			None
R_S increased			
L decreased			
L increased			
L open			
R_S open			

Questions

1. Explain the variations observed in Step 4.

2. Explain the variations observed in Step 6.

3. Explain the variations observed in Step 8.

4. Explain the variations observed in Step 10.

5. Explain the variations observed in Step 12.

6. When the 100 mH inductor was inserted, was the measured phase angle more or less ideal? Why?

Exercise 14

Transformers

OBJECTIVES

After completing this exercise, you should be able to:

- Determine the turns ratio of a transformer.
- Determine if a transformer has a step-up or a step-down configuration.
- Determine the reflected load impedance in the primary of a transformer.
- Demonstrate the phase relationship between the outputs of a center-tapped transformer.

DISCUSSION

A transformer uses *electromagnetic induction* to couple an ac signal from its input (primary) to its output (secondary). Depending on the *turns ratio* between the primary and the secondary windings, a transformer may step up or step down an input voltage. Despite occasionally-significant differences in primary and secondary voltage, the power in the primary circuit is approximately equal to the power in the secondary circuit. If not for losses in the transformer, they would be (ideally) equal. In this exercise, you will substantiate this concept.

Later in your studies, you will learn about power supplies. In some power supplies, *center-tapped* transformers play a significant role. The phase relationship at the output of a center-tapped transformer is critical to the function of any power supply using this component. At the end of this exercise, you will briefly examine the output characteristics of a center-tapped transformer.

LAB PREPARATION

Review Section 10.5 of *Electronics Technology Fundamentals*.

MATERIALS

    1   dual-trace oscilloscope

    1   DMM

    1   function generator

    1   protoboard

    2   100 Ω resistors

    1   120 V/25.2 V center-tapped transformer

PROCEDURE

1. Construct the circuit shown in Figure 14.1. Connect Channel 1 (C1) of the oscilloscope to measure the voltage across the transformer primary. Set the output of the signal generator for a 10 V_{PP} sine wave at a frequency of 60 Hz. Use Channel 2 (C2) to measure the voltage across the transformer load (R_L). Draw the transformer input and output waveforms in Figure 14.2, and record the peak-to-peak readings of both.

$$V_P = \underline{\hspace{1.5cm}} \qquad\qquad V_{RL} = \underline{\hspace{1.5cm}}$$

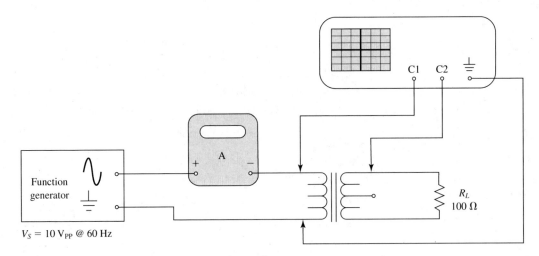

FIGURE 14.1

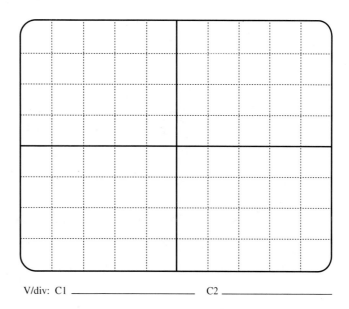

V/div: C1 _____ C2 _____

Time Base _____

FIGURE 14.2 Input and output waveforms for the circuit shown in Figure 14.1.

2. Use the primary and secondary voltages measured in Step 1 to calculate the turns ratio of the transformer.

$$\frac{N_P}{N_S} = \frac{V_P}{V_S} = \underline{\hspace{2cm}}$$

3. Use the ammeter (set to ac current) to measure the primary current, and convert the primary voltage to an rms value. (*Remember:* AC meters always measure rms values.) Use the rms current and voltage to calculate the power in the primary circuit.

$$I_P = \underline{\hspace{2cm}} \text{ (rms)}$$

$$V_P = \underline{\hspace{2cm}} \text{ (rms)}$$

$$P_P = \underline{\hspace{2cm}}$$

4. Convert the secondary voltage that you measured in Step 1 to an rms value. Use this value and the load resistance to calculate the power in the secondary circuit.

$$V_S = \underline{\hspace{2cm}} \text{ (rms)}$$

$$P_S = \underline{\hspace{2cm}}$$

5. Use your results from Step 3 to calculate the impedance of the primary. (Be sure to use the rms values.)

$$Z_P = \underline{\hspace{2cm}}$$

6. Now, calculate the impedance in the primary using the relationship below. Assume that $Z_S = R_L$.

$$Z_P = Z_S \left(\frac{N_P}{N_S}\right)^2 = \underline{\hspace{2cm}}$$

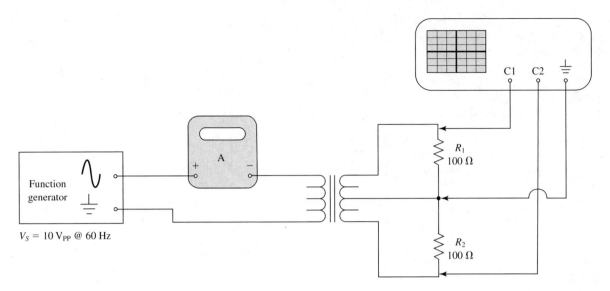

FIGURE 14.3

7. Construct the circuit shown in Figure 14.3.
8. Before connecting the oscilloscope to the secondary, temporarily connect Channel 1 to measure the output from the function generator. Set the generator output signal to 10 V_{PP} at 60 Hz. After setting the transformer input signal, connect both scope channels to the secondary as shown in Figure 14.3.
9. Set the scope to trigger on Channel 1. Measure and compare the voltage across R_1 and R_2, and note their phase relationship.

$$V_{R1} = \underline{\hspace{3cm}}$$

$$V_{R2} = \underline{\hspace{3cm}}$$

$$\theta = \underline{\hspace{3cm}}$$

QUESTIONS & PROBLEMS

1. Is the transformer used in this exercise a step-up or a step-down transformer? Explain your answer.

2. Refer to your waveforms in Figure 14.2. Describe the phase relationship between these two waveforms.

3. The transformer that you used for this exercise is rated as a 120 V/25.2 V center-tapped transformer. Explain how your results from Step 2 substantiate (or refute) this rating.

4. Refer to your results from Steps 3 and 4. Were the input and output power values equal? Explain any discrepancy.

5. Refer to your results from Steps 5 and 6. Did the results of your calculations for primary impedance agree? Explain any variation between the two results.

6. Refer to your results from Step 9. Explain the phase relationship between V_{R1} and V_{R2}. (*Hint:* It relates to the center tap.)

SIMULATION EXERCISE

Procedure

1. Open file Ex14.1 from the Electronics Technology Fundamentals companion web site (www.prenhall.com/paynter). Run the simulation and record the measured current and voltage values below. These are the "normal" circuit values.

$I_1 =$ _____ $I_2 =$ _____ $I_3 =$ _____

$V_{R1} =$ _____ $V_{R2} =$ _____ $V_{R3} =$ _____

Fault Analysis

2. Insert each fault listed in Table 14.1 and run the simulation with the fault present. Compare the indications with those of the "normal" circuit, and note any differences in the space provided. Remember to return each component to its original condition before inserting the new fault.

Questions

1. Do the current readings verify Kirchhoff's current law in the reference circuit? Explain your answer.

TABLE 14.1   Simulation Faults and Readings

Fault	Resulting Changes in Current and Voltage Readings
Open primary (terminal 1)	
Open secondary (top, terminal 3)	
Open secondary (center-tap, terminal 5)	
Open secondary (bottom, terminal 4)	
Shorted 1 kΩ resistor	
Shorted 2 kΩ resistor	
Shorted 3 kΩ resistor	

2. Do the voltage readings verify Kirchhoff's voltage law in the reference circuit? Explain your answer.

3. What is the full secondary voltage of this transformer? Is this transformer acting as a step-up or a step-down transformer?

4. Explain the readings that resulted when the transformer primary was faulted open.

5. When transformer terminal 3 was faulted open, the ammeter closest to the fault indicated zero. Explain why all the other meters were indicating a value other than 0 A. Use Kirchhoff's current law to justify your answer.

6. When transformer terminal 5 was faulted open, the ammeter closest to the fault indicated zero. Explain why all the other meters were indicating a value other than 0 A. Use Kirchhoff's current law to justify your answer.

7. Why did terminal 5 being open give different results than terminal 3 being open?

8. Why did the voltages change so drastically when the transformer center tap (terminal 5) opened?

9. What was the result of any short in the secondary circuit? Explain why this happened, given that there are two separate circuits (primary to secondary).

Exercise 15

Series RL Circuits

OBJECTIVES

After completing this exercise, you should be able to:

- Determine the impedance of a series *RL* circuit.
- Demonstrate the effect that changes in frequency have on a series *RL* circuit.
- Demonstrate that a series *RL* circuit complies with the basic principles of series circuits.

DISCUSSION

Series *RL* circuits comply with all the basic principles of series circuits in general. That is:

- The current is equal at all points in the circuit.
- The applied voltage equals the sum of the component voltages.
- The total opposition to current equals the sum of the component oppositions.

However, the phase angles that exist between the component values affect the approach that you use to analyze and describe the circuit. You will examine the effect that phase angles have on circuit analysis in this exercise. You will also observe the effects that changes in frequency have on the operation of series *RL* circuits.

One other point must be made. In the text, you were introduced to the idea of using current as the 0° reference for phase angle measurements (since current is equal at all points in the circuit). Since resistor voltage (V_R) is in phase with the circuit current, you could use

V_R as the 0° reference. Even so, you will be required (at specific points in the exercise) to swap the inductor and resistor positions. This affects your ability to visually compare the various circuit waveforms to the resistor voltage on the oscilloscope screen. For this reason, V_S *will be used as the reference for your circuit phase angles.* When the oscilloscope is connected as shown in Figures 15.2 and 15.3, Channel 1 displays the phase reference (V_S). An example of a phasor diagram that is referenced to V_S is shown in Figure 15.1. Note that the 90° phase angle between V_R and V_L is represented in the figure. Once all the phase angles are established and the phasors are drawn, you will then alter them to match the form discussed in the text.

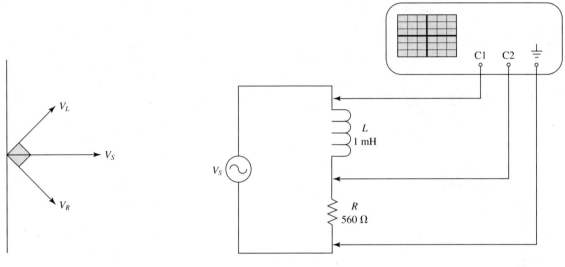

FIGURE 15.1 Inductive and resistive phasors referenced to source voltage.

FIGURE 15.2

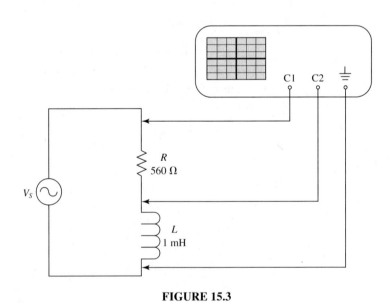

FIGURE 15.3

LAB PREPARATION

Review Section 11.1 of *Electronics Technology Fundamentals.*

MATERIALS

1   dual-trace oscilloscope

1   function generator

1   protoboard

1   560 Ω resistor

1   1 mH inductor

PROCEDURE

1. Construct the circuit shown in Figure 15.2. Channel 1 (C1) is connected to display (and trigger on) V_S. Channel 2 (C2) is connected to display V_R.
2. Set the function generator for an 8 V_{PP} sine wave at a frequency of 20 kHz.
3. Measure the peak-to-peak value of V_R and record your measurement in Table 15.1.

TABLE 15.1   Measurements for the Series *RL* Circuit

Frequency	V_R	θ_R	V_L	θ_L
20 kHz				
45 kHz				
90 kHz				
180 kHz				
350 kHz				

4. The phase angle of V_R relative to V_S can be determined using

$$\theta = (360°)\frac{t}{T}$$

where

T = the waveform period

t = the time between the points where the two waveforms cross the center line on the oscilloscope grid

These times are illustrated in Figure 13.4 (page 100 of this manual). Determine the phase angle of V_R relative to V_S, and record this value in Table 15.1. (*Remember:* If V_R *lags* V_S, the phase angle is *negative*. If V_R *leads* V_S, the phase angle is *positive*.)

5. Repeat Steps 3 and 4 at the remaining frequencies listed in Table 15.1. Make certain that V_S remains constant at 8 V_{PP} as you change frequencies.
6. Reverse the components as shown in Figure 15.3, and repeat Steps 1 through 5 for this circuit. This time, you are measuring V_L and its phase angle relative to V_S. Enter your results in Table 15.1.

7. Using your data from Table 15.1, calculate the magnitude and phase angle of V_S, and enter your results in Table 15.2.

TABLE 15.2 Calculations for the Series *RL* Circuit

Frequency	$V_S = \sqrt{V_R^2 + V_L^2}$	$\theta = \tan^{-1}\left(\dfrac{V_L}{V_R}\right)$
20 kHz		
45 kHz		
90 kHz		
180 kHz		
350 kHz		

8. Use your results from Tables 15.1 and 15.2 to draw the voltage phasor diagrams for this circuit at frequencies of 20 kHz, 90 kHz, and 350 kHz in Figure 15.4. Remember, at this point V_S is your phase reference.

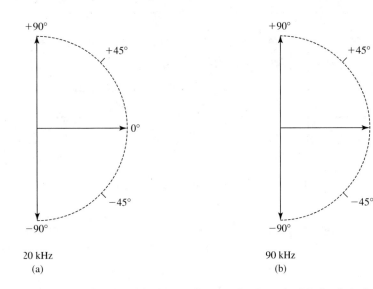
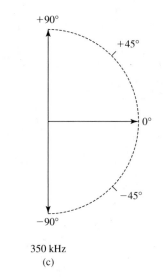

| 20 kHz | 90 kHz | 350 kHz |
| (a) | (b) | (c) |

FIGURE 15.4 Phasor diagrams for the series *RL* circuit (referenced to V_S) at three operating frequencies.

9. Alter your phasor diagram to match the form introduced in the text as follows:
   a. Determine the value that, when added to the phase angle of V_R, results in a phase angle of 0°.
   b. Add the value from Step 9a to the voltage phase angles.
   c. Draw the new phasors in Figure 15.5.

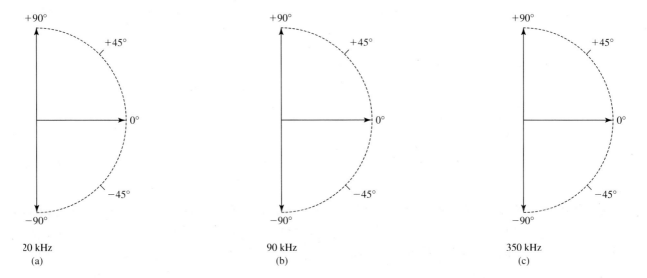

20 kHz
(a)

90 kHz
(b)

350 kHz
(c)

FIGURE 15.5 Phasor diagrams for the series RL circuit (referenced to V_R) at three operating frequencies.

10. Calculate X_L for each frequency shown in Table 15.3. Use these values, your results from Table 15.1, and Ohm's law to calculate I_L. Enter those values in Table 15.3. (*Remember:* X_L has a phase angle of 90°.)

TABLE 15.3

Frequency	X_L	I_L	R	I_R
20 kHz			560 Ω	
45 kHz			560 Ω	
90 kHz			560 Ω	
180 kHz			560 Ω	
350 kHz			560 Ω	

11. Use your results from Table 15.1 and Ohm's law to calculate I_R for each frequency shown in Table 15.3, and enter your results in that table. (*Remember:* R has a phase angle of 0°.)

QUESTIONS & PROBLEMS

1. Refer to Table 15.1. As frequency increased, did the circuit become more or less inductive? Explain your answer.

2. Refer to Table 15.2. Explain how these results prove that Kirchhoff's voltage law applies to series RL circuits.

3. Refer to Table 15.3. Based on your current values, what happens to circuit impedance when frequency increases in a series RL circuit? Explain your answer.

4. Use the values of X_L and R from Table 15.3 to calculate Z_T and its phase angle. Enter these values in Table 15.4, and plot a curve representing the relationship between the impedance phase angle and frequency in Figure 15.6.

TABLE 15.4

Frequency	$Z_T = \sqrt{R^2 + X_L^2}$	$\theta = \tan^{-1}\left(\dfrac{X_L}{R}\right)$
20 kHz		
45 kHz		
90 kHz		
180 kHz		
350 kHz		

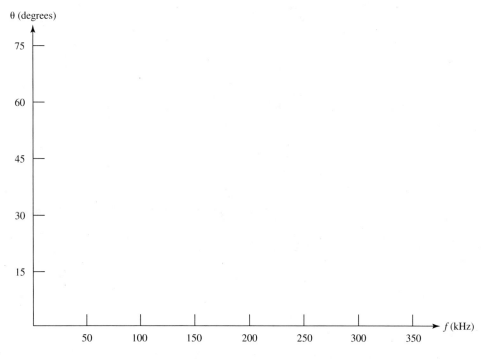

θ (degrees)

FIGURE 15.6 Impedance phase angles versus frequency.

SIMULATION EXERCISE

Procedure

1. Open file Ex15.1 from the Electronics Technology Fundamentals companion web site (www.prenhall.com/paynter). This is the same circuit as the one shown in Figure 15.2.
2. Run the simulation. Measure and record the phase angle of V_R relative to V_S in Table 15.5.

TABLE 15.5

Condition	θ @ 20 kHz	f(θ = 45°)	V_R @ f	X_L @ f
R = 560 Ω				
R = 56 Ω				
R = 5.6 kΩ				

3. With the source voltage constant at 8 V_{PP}, vary the frequency until the phase angle is approximately 45°. Resimulate as necessary. Measure and record the peak-to-peak value of V_R and the frequency at which $\theta = 45°$.
4. Change the value of the resistor to 56 Ω, and reset the frequency to 20 kHz.
5. Repeat Steps 2 and 3.
6. Change the value of the resistor to 5.6 kΩ, and reset the frequency to 20 kHz.
7. Repeat Steps 2 and 3.
8. Calculate X_L at each value of f in Table 15.5.

Questions

1. What happened to the $\theta = 45°$ frequency as the resistor value changed? Why do you think these changes occurred?

2. Compare the calculated values of X_L in Table 15.5 with their associated resistance values. Based on your observations, what is the relationship between R and X_L when $\theta = 45°$?

Exercise 16

Parallel RL Circuits

OBJECTIVES

After completing this exercise, you should be able to:

- Determine the impedance of a parallel *RL* circuit.
- Demonstrate the effect that changes in frequency have on a parallel *RL* circuit.
- Demonstrate that a parallel *RL* circuit complies with all the basic rules of parallel circuits.

DISCUSSION

As in any parallel circuit, Kirchhoff's current law tells us that the sum of the branch currents in a parallel *RL* circuit must equal the total current entering the circuit. As you know, inductor current *lags* inductor voltage. Because of the relative phase angle between the inductor current and the current through the resistor, the two currents must be added *geometrically.*

One problem in Exercise 15 was that source voltage had to be used as a reference. Normally, in a series circuit, current is used as the 0° phase-angle reference. In this exercise, you will not have this problem. Because the voltage across a parallel circuit is common for all the branches in the network, voltage is used as the phase reference.

In this exercise, you are once again using a current sensing resistor (R_S) to help measure the phase angle of I_T. Since resistive current and voltage are in phase, the phase angle of V_{RS} equals the phase angle of I_T. Thus, by viewing V_{RS} and V_S simultaneously on the oscilloscope, you can see the phase angle of I_T with respect to V_S.

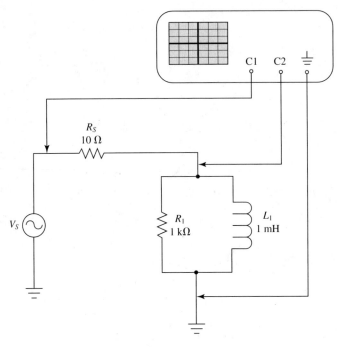

FIGURE 16.1

LAB PREPARATION

Review Section 11.3 of *Electronics Technology Fundamentals*.

MATERIALS

1   dual-trace oscilloscope

1   function generator

1   protoboard

2   resistors: 10 Ω and 1 kΩ

1   1 mH inductor

PROCEDURE

1. Construct the circuit shown in Figure 16.1. Channel 1 (C1) is connected to display V_S. Channel 2 (C2) is connected to display the V_{R1} and V_L waveforms. Note that the voltage across the sensing resistor (V_{RS}) will be used to calculate I_T and so observe the phase angle of I_T relative to V_S.

2. Set the function generator for a 10 V_{PP} sine wave at a frequency of 20 kHz. Measure the peak-to-peak voltage across the parallel network (V_{NET}), and record your measurement in Table 16.1. Using the technique introduced in Exercise 15 (step 4, p. 119), determine the phase angle of the voltage across the parallel circuit relative to V_S. Record this value in the second column of Table 16.1.

TABLE 16.1   Measured Values for the Circuit Shown in Figure 16.1

Frequency	V_{NET}	$\theta\,(V_{NET})$	V_{RS}	$\theta\,(V_{RS})$
20 kHz				
40 kHz				
80 kHz				
160 kHz				
320 kHz				
640 kHz				
1.2 MHz				

3. Repeat Step 2 for the remaining frequencies listed in Table 16.1. Make certain that V_S remains constant at 10 V_{PP} as you change frequencies.

4. Reverse the components as shown in Figure 16.2, and repeat Steps 1 through 3 for this circuit. This time, you are measuring V_{RS} and its phase angle relative to V_S. Enter your results in the two right-hand columns of Table 16.1.

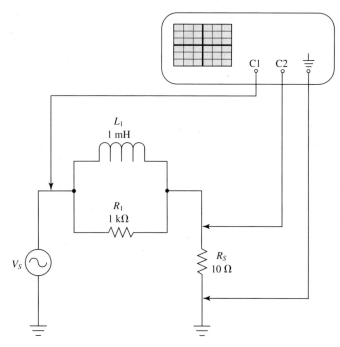

FIGURE 16.2

5. Calculate X_L for each frequency in Table 16.1, and record these values in Table 16.2. Use the values for V_{NET} and X_L to calculate the magnitude and phase angle of I_L for each frequency. Enter these results in Table 16.2.

TABLE 16.2

Frequency	X_L	I_L	I_{R1}	I_T	$I_T = \sqrt{I_L^2 + I_{R1}^2}$	$\theta = \tan^{-1}\left(\dfrac{-I_L}{I_{R1}}\right)$
20 kHz						
40 kHz						
80 kHz						
160 kHz						
320 kHz						
640 kHz						
1.2 MHz						

6. Use the values of V_{NET} and R_1 to calculate the magnitude and phase angle of I_{R1} for each frequency. Enter your results in Table 16.2.
7. Use the values of V_{RS} and R_S to calculate the magnitude and phase angle of I_T for each frequency. Enter your results in Table 16.2.
8. Use the values of I_L and I_{R1} to calculate the magnitude and phase angle of I_T. Enter your results in Table 16.2.
9. Use your results from Table 16.2 to draw the current phasor diagrams for this circuit at frequencies of 20 kHz, 160 kHz, and 1.2 MHz in Figure 16.3.

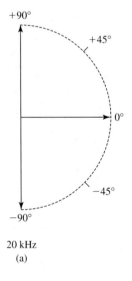

20 kHz
(a)

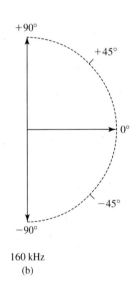

160 kHz
(b)

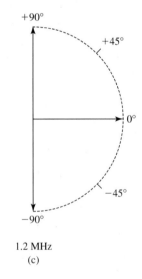

1.2 MHz
(c)

FIGURE 16.3

10. Use your results from Table 16.2 to calculate the magnitude and phase angle of Z_T for each frequency listed in Table 16.3.
11. Use your results from Table 16.2 to draw a curve representing the relationship between the current phase angle and frequency in Figure 16.4a.
12. Use your results from Table 16.3 to draw a curve representing the relationship between the impedance phase angle and frequency in Figure 16.4b.

TABLE 16.3

Frequency	$Z_T \angle \theta$
20 kHz	
40 kHz	
80 kHz	
160 kHz	
320 kHz	
640 kHz	
1.2 MHz	

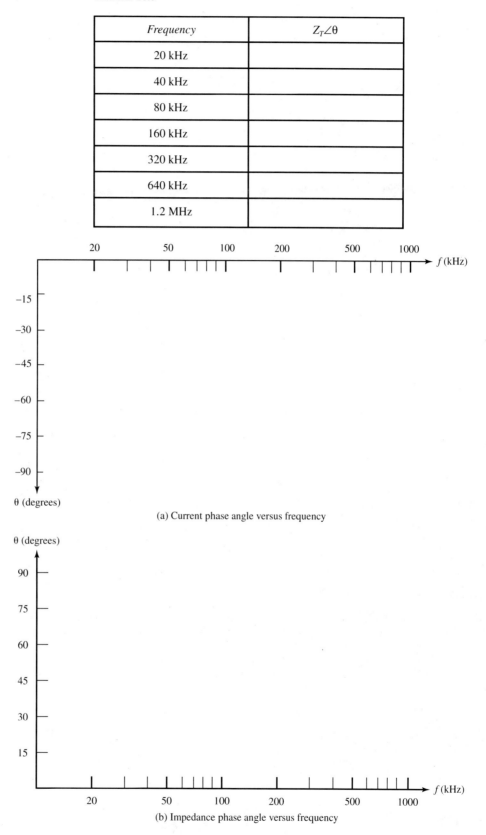

(a) Current phase angle versus frequency

(b) Impedance phase angle versus frequency

FIGURE 16.4

QUESTIONS & PROBLEMS

1. Refer to Table 16.1. As the frequency increased, did the total circuit current become more or less inductive? Explain your answer.

2. Refer to Table 16.2. Explain how your results prove that Kirchhoff's current law holds true for parallel *RL* circuits.

3. Refer to your phasor diagrams in Step 9. Explain what happens to the two branch currents in a parallel *RL* circuit when the frequency changes. Be sure to comment on their respective phase angles.

4. Refer to the phase angle curves that you generated in Steps 11 and 12. Explain why the phase angle for I_T becomes less negative as the frequency increases, while the phase angle for Z_T becomes less positive. (*Hint:* Ohm's law might be a good place to start.)

Discussion

A simulator allows a circuit to be "tested" before building it in lab, or the results of a lab activity to be checked. It is sometimes beneficial to run the exercise with a different set of instruments to verify that the results are valid even though the instruments provide different numbers.

In practice, many ac current meters are limited in the results they can provide due to their limited frequency response and/or their internal resistance. An ac meter provides an *effective* (rms) value of voltage or current, whereas the scope provides either peak or peak-to-peak voltage values. Comparing the results from an *rms indicator* with a *peak* or a *peak-to-peak indicator* can provide both a cross-check of results and an opportunity to practice ac value conversions.

Procedure

1. Open file Ex16.1 from the Electronics Technology Fundamentals companion web site (www.prenhall.com/paynter). This is the same circuit as the one shown in Figure 16.1. Note that multimeters are being used as ac ammeters since they provide better resolution than current indicators.
2. Run the simulation for each frequency listed in Table 16.4, and record the values of I_T, I_R, and I_L in the table.

TABLE 16.4   Simulation Results and Calculations

Frequency	I_T	I_R	I_L	$\sqrt{I_L^2 + I_R^2}$	$\tan^{-1}\left(\dfrac{-I_L}{I_R}\right)$
20 kHz					
40 kHz					
80 kHz					
160 kHz					
320 kHz					
1.2 MHz					

3. Using your simulation results, complete the last two columns in Table 16.4.

Questions

1. Do the measured and calculated values of I_T in Table 16.4 closely match each other? Explain any discrepancies in the two values.

2. Can the currents in Table 16.4 be directly compared with those in Table 16.2? If not, what must be done to allow the comparison?

3. How do the simulator phase angles compare with those determined in the hardware lab procedure?

4. If the value of the current-sensing resistor shown in Figure 16.1 (which is in the position traditionally used to represent source resistance) is increased, what effect would the change have on the impedance and phase angle "seen" by the source?

Exercise 17

Capacitors and Capacitive Reactance (X_C)

OBJECTIVES

After completing this exercise, you should be able to:

- Explain the relationship between voltage and current in a capacitive circuit.
- Explain the relationship between capacitive reactance and frequency.
- Determine capacitive reactance using Ohm's law.

DISCUSSION

A capacitor acts, in many ways, like the mirror image of an inductor. In the first section of this exercise, you will verify that capacitor voltage *lags* capacitor current by 90°, as illustrated in Figure 17.1. To observe this phase relationship, you will once again employ a

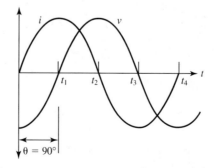

FIGURE 17.1   The phase relationship between capacitor voltage and current.

sensing resistor to produce a voltage that is in phase with the circuit current. This voltage can then be used to verify the phase relationship between capacitor current and voltage.

In the second part of this exercise, you will verify the relationship between capacitive reactance (X_C) and frequency. As you will see, X_C varies *inversely* with frequency.

LAB PREPARATION

Review Sections 12.1, 12.2, and 12.4 of *Electronics Technology Fundamentals*.

MATERIALS

1   dual-trace oscilloscope
1   function generator
1   variable dc power supply
1   protoboard
2   resistors: 1 kΩ and 100 Ω
1   10 nF capacitor

PROCEDURE

Part 1: The Phase Relationship Between Capacitor Current and Voltage

1. Construct the circuit shown in Figure 17.2. Note the following:
   a. Connecting Channel 1 (C1) of the oscilloscope to point A allows you to monitor the voltage across the entire circuit. Since $R_S \ll X_C$ at the circuit operating frequency, $V_{RS} \ll V_C$. Therefore, the waveform displayed on Channel 1 can be assumed to represent V_C.
   b. Connecting Channel 2 (C2) of the oscilloscope to point B allows you to measure the voltage across the sensing resistor. Because the resistor voltage and current are in phase, V_{RS} is in phase with the capacitor current. As a result, the oscilloscope waveforms represent the phase relationship between V_C (Channel 1) and I_C (Channel 2).

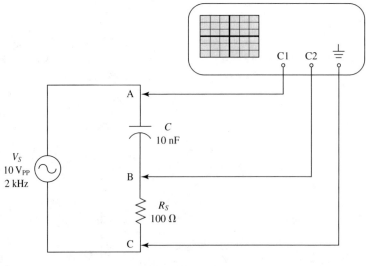

FIGURE 17.2

2. Set the signal generator for a 10 V$_{PP}$ sine wave at a frequency of 2 kHz. Adjust the vertical sensitivity *(volts/div)* setting for Channel 2 so that the two waveform displays are approximately equal in size. (They should appear similar to those shown in Figure 17.1.)
3. Draw the two waveforms to scale in Figure 17.3.

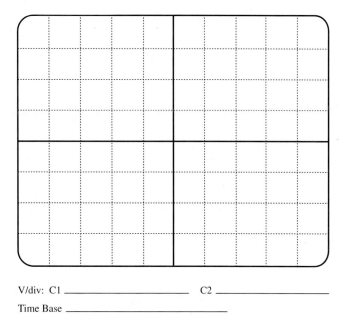

V/div: C1 _____ C2 _____

Time Base _____

FIGURE 17.3   Waveforms for the circuit in Figure 17.2.

Part 2: Capacitive Reactance and Frequency

4. Construct the circuit shown in Figure 17.4. Set the output of the signal generator for an 8 V$_{PP}$ sine wave at a frequency of 2 kHz. Use Channel 1 of the oscilloscope to display the output of the generator, and use Channel 2 to display the voltage across R_1.

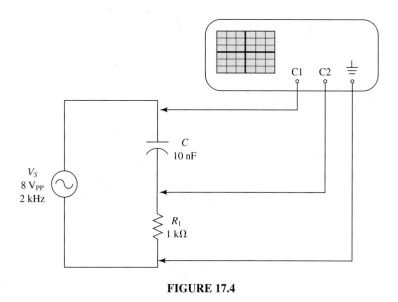

FIGURE 17.4

5. Measure the peak-to-peak value of V_{R1}. Record your measurement in Table 17.1.

TABLE 17.1

Frequency	V_{R1}	I_T	V_C	$X_C = \dfrac{V_C}{I_T}$	$X_C = \dfrac{1}{2\pi f C}$
2 kHz					
5 kHz					
10 kHz					
15 kHz					
25 kHz					
75 kHz					
150 kHz					

6. Repeat Step 5 for frequencies of 5 kHz, 10 kHz, 15 kHz, 25 kHz, 75 kHz, and 150 kHz. Make certain that the signal applied to the circuit remains at 8 V_{PP} when you change frequencies. Enter your results in Table 17.1.
7. Reverse the position of the components as shown in Figure 17.5 so that Channel 2 displays the voltage across the capacitor. Repeat Steps 4, 5, and 6, and record the values of V_C in Table 17.1.

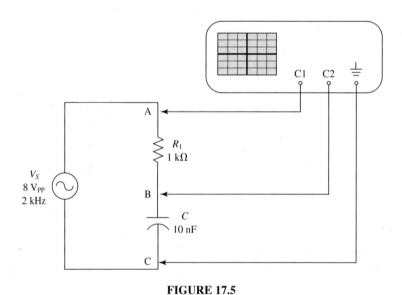

FIGURE 17.5

8. Using the values of V_{R1} and R_1, calculate I_T for each frequency. Enter your calculations in Table 17.1.
9. Using the values of I_T and V_C from Table 17.1, calculate X_C. Enter your calculations in Table 17.1.
10. Using the frequencies listed in Table 17.1, complete the right-hand column in that table.

11. Using the values in the right-hand column in Table 17.1, plot a curve representing the relationship between capacitive reactance and frequency in Figure 17.6.

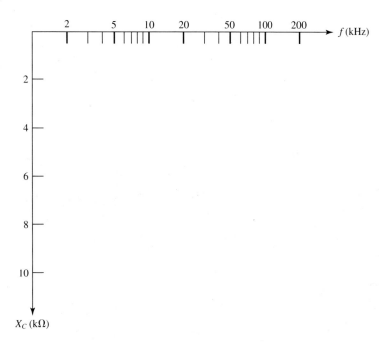

FIGURE 17.6

QUESTIONS & PROBLEMS

1. Refer to the waveforms drawn in Figure 17.3. Do the waveforms appear to be exactly 90° out of phase? If not, explain any discrepancies.

2. Explain how you can tell from the results of Step 3 that current *leads* voltage, rather than the other way around.

3. Refer to Table 17.1 and the curve in Figure 17.6. Explain what these results tell you about the relationship between frequency and capacitive reactance.

4. Refer to Table 17.1. Did the two methods of calculating X_C agree with each other? If not, explain any discrepancies.

5. Refer to Table 17.1. Explain why circuit current increased as frequency increased.

SIMULATION EXERCISE

Procedure

1. Open file Ex17.1 from the Electronics Technology Fundamentals companion web site (www.prenhall.com/paynter). This is the same circuit as the one shown in Figure 17.2.
2. Measure the peak-to-peak value of V_R and the phase angle of V_R relative to V_S. These measurements represent the "normal" operation of the circuit. Record these values in Table 17.2.

TABLE 17.2

Condition	V_R	θ	Variation from Normal Operation
Normal operation			None
R_S increased (×10)			
C increased (×10)			
C decreased (×0.1)			
C open			
R_S open			

3. Fault the circuit by changing R_S to 1 kΩ. Run the simulation.
4. Measure the peak-to-peak value of V_R and the phase angle of V_R relative to V_S. Record these values in Table 17.2.
5. Note and record the changes that occur as a result of the fault.
6. Restore R_S to its initial value, and fault the circuit by changing the capacitor value to 100 nF. Run the simulation.
7. Repeat Steps 4 and 5.
8. Change the capacitor value to 1 nF, and run the simulation.
9. Repeat Steps 4 and 5.
10. Fault the capacitor open, and run the simulation.
11. Repeat Steps 4 and 5.
12. Restore the capacitor to its initial value, and fault R_S open. Run the simulation.
13. Repeat Steps 4 and 5.

Questions

1. What effect did increasing R_S by a factor of 10 have on the values of V_R and θ?

2. What effect did increasing C by a factor of 10 have on the values of V_R and θ?

3. Compare your responses to Questions 1 and 2. How do the changes noted compare with each other?

4. When $C = 1$ nF, is the measured phase angle more or less ideal? Why?

5. Explain the variations noted in Table 17.2 when the capacitor is open.

6. Explain the variations noted in Table 17.2 when the resistor is open.

Exercise 18

Series RC Circuits

After completing this exercise, you should be able to:

- Determine the impedance of a series *RC* circuit.
- Demonstrate the effect that a change in frequency has on a series *RC* circuit.
- Demonstrate that the basic rules of series circuits are fulfilled by a series *RC* circuit.

DISCUSSION

In Exercise 15, you looked at the characteristics of series *RL* circuits. In this exercise, you will examine the characteristics of series *RC* circuits. Like series *RL* circuits, series *RC* circuits comply with all the basic rules of series circuit operation:

- Current is the same at any point in the circuit.
- The sum of the component voltages must equal the source voltage.
- The total opposition to current is equal to the sum of the component oppositions.

As you saw in Exercise 15, however, the phase relationship between current and voltage must be considered when solving for currents and voltages in a series resistive-reactive circuit.

Another point must be made: In the text, you were introduced to the idea of using current as the 0° reference for phase angle measurements, since current is equal at all points in a series circuit. Resistor voltage (V_R) is in phase with the circuit current, so we could use V_R as the 0° reference. At specific points in this exercise, however, you will be required to swap the capacitor and resistor positions. Since this affects our ability to visually compare the various circuit waveforms to the resistor voltage on the oscilloscope screen, V_S *will be used as the reference for our circuit phase angles*. As a result, the phasor diagram is referenced to

V_S, as shown in Figure 18.1. Note that the 90° phase angle between V_R and V_C is represented in the figure. Once all the phase angles are established and the phasors are drawn, you will then alter them to match the form discussed in the text.

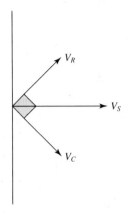

FIGURE 18.1

LAB PREPARATION

Review Section 13.1 of *Electronics Technology Fundamentals*.

MATERIALS

1   dual-trace oscilloscope
1   function generator
1   protoboard
1   330 Ω resistor
1   0.1 μF capacitor

PROCEDURE

1. Construct the circuit shown in Figure 18.2. Channel 1 (C1) is connected across the entire circuit and displays V_S. Channel 2 (C2) is connected to display V_R. Set the function generator for an 8 V_{PP} sine wave at a frequency of 1 kHz.

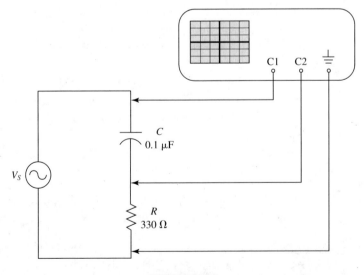

FIGURE 18.2

2. Measure the peak-to-peak value of V_R and record your measurement in Table 18.1. Measure the phase angle of V_R relative to V_S. Record this value (as θ_R) in Table 18.1.

TABLE 18.1

Frequency	V_R	θ_R	V_C	θ_C
1 kHz				
2 kHz				
5 kHz				
10 kHz				
20 kHz				

3. Repeat Step 2 for the other frequencies listed in Table 18.1. Make certain that V_S remains constant at 8 V_{PP} as you change frequencies.
4. Reverse the components as shown in Figure 18.3. Repeat Steps 1 through 3 for this circuit, measuring V_C and its phase angle relative to V_S. Enter your results in the two right-hand columns of Table 18.1.

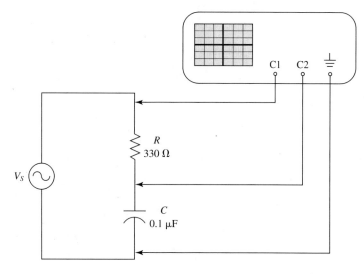

FIGURE 18.3

5. Use the values for V_R and V_C in Table 18.1 to calculate the magnitude and phase angle of V_S. Enter your results in Table 18.2.

TABLE 18.2

Frequency	V_S	θ
1 kHz		
2 kHz		
5 kHz		
10 kHz		
20 kHz		

6. Use your results from Table 18.1 to draw the voltage phasor diagrams for V_R and V_C at frequencies of 1 kHz, 5 kHz, and 20 kHz in Figure 18.4.

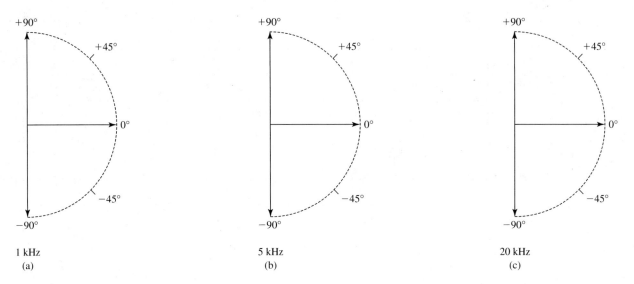

FIGURE 18.4 Phasor diagrams for the series *RC* circuit (referenced to V_S) at three operating frequencies.

7. Alter your phasor diagram to match the form introduced in the text, as follows:
   a. Determine the value that, when added to the phase angle of V_R, results in a phase angle of 0°.
   b. Add the value from Step 7a to the voltage phase angles.
   c. Draw the new phasors in Figure 18.5.

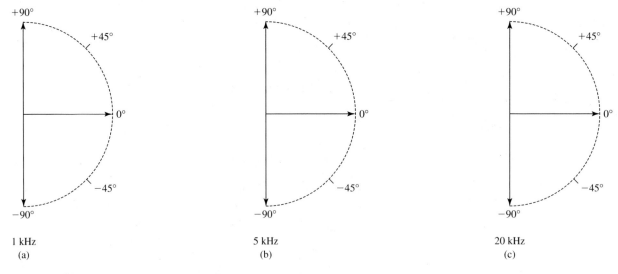

FIGURE 18.5 Phasor diagrams for the series *RC* circuit (referenced to circuit current) at three operating frequencies.

8. Calculate X_C for each frequency shown in Table 18.3. Use these values and your results from Table 18.1 to calculate I_C. Enter these values in Table 18.3. (*Remember:* X_C has a phase angle of −90°.)

TABLE 18.3

Frequency	X_C	I_C	R	I_R
1 kHz			330 Ω	
2 kHz			330 Ω	
5 kHz			330 Ω	
10 kHz			330 Ω	
20 kHz			330 Ω	

9. Use your results from Table 18.1 to calculate I_R for each frequency shown in Table 18.3. Enter your results in the table. (*Remember: R* has a phase angle of 0°.)

QUESTIONS & PROBLEMS

1. Refer to your results in Table 18.1. As frequency increased, did the circuit become more or less capacitive? Explain your answer.

2. Refer to your results in Table 18.2. Explain how your results demonstrate that Kirchhoff's voltage law applies to series *RC* circuits.

3. Refer to your current calculations in Table 18.3. What basic rule of series circuits do these results support? Explain your answer.

4. Use the values of X_C from Table 18.3 to calculate Z_T and its phase angle. Enter these values in Table 18.4. Then, plot a curve representing the relationship between the impedance phase angle and frequency in Figure 18.6.

TABLE 18.4

Frequency	Z_T	θ
1 kHz		
2 kHz		
5 kHz		
10 kHz		
20 kHz		

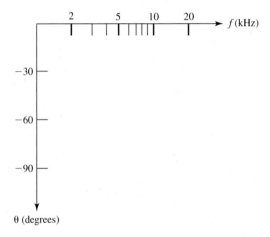

FIGURE 18.6

SIMULATION EXERCISE

Procedure

1. Open file Ex18.1 from the Electronics Technology Fundamentals companion web site (www.prenhall.com/paynter). This is the same circuit as the one shown in Figure 18.2.
2. Run the simulation. Measure and record the phase angle of V_R relative to V_S in Table 18.5. Note that this represents the phase angle of I_T with respect to V_S.

TABLE 18.5

Condition	θ @ 1 kHz	$f(θ = 45°)$	V_R @ f	X_C @ f
$R = 330\ \Omega$				
$R = 33\ \Omega$				
$R = 3.3\ k\Omega$				

3. With the source voltage kept constant at 8 V_{PP}, vary the frequency until the phase angle is approximately 45°. Resimulate as necessary. Measure and record the peak-to-peak value of V_R and the frequency at which $\theta = 45°$. Record these values in Table 18.5.
4. Change the value of the resistor to 33 Ω, and reset the frequency to 1 kHz.
5. Repeat Steps 2 and 3.
6. Change the value of the resistor to 3.3 kΩ, and reset the frequency to 1 kHz.
7. Repeat Steps 2 and 3.
8. Calculate the value of X_C at each value of f in Table 18.5.

Questions

1. What happened to the $\theta = 45°$ frequency as the resistor value changed? Why do you think these changes occurred?

2. Compare the calculated values of X_C in Table 18.5 with their associated resistance values. Based on your observations, what is the relationship between R and X_C when $\theta = 45°$?

Exercise 19

Parallel RC Circuits

OBJECTIVES

After completing this exercise, you should be able to:

- Determine the impedance of a parallel *RC* circuit.
- Demonstrate the effect that a change in frequency has on a parallel *RC* circuit.
- Demonstrate that a parallel *RC* circuit complies with the basic rules of parallel circuits.

DISCUSSION

Parallel *RC* circuits comply with all the basic rules of parallel circuits:

- The branch voltages are equal, regardless of whether a resistor or a capacitor is in the branch.
- The sum of the branch currents equals the total circuit current.

However, since the resistive and capacitive currents are out of phase, you must consider the phase angles when calculating the total circuit current.

In Exercise 17, you observed that capacitor current *leads* capacitor voltage by 90°. Since the voltage across the capacitive and resistive branches in a parallel *RC* circuit are equal, the current through the capacitive branch leads the current through the resistive branch by 90°. Thus, when you add the two branch currents, you must add them *geometrically*.

LAB PREPARATION

Review Section 13.3 of *Electronics Technology Fundamentals*.

MATERIALS

1   dual-trace oscilloscope

1   function generator

1   protoboard

2   resistors: 47 Ω and 1.5 kΩ

1   10 nF capacitor

PROCEDURE

1. Construct the circuit shown in Figure 19.1. Channel 1 (C1) is connected to display the V_S waveform. Channel 2 (C2) is connected to display V_{R1} and V_C. Note the *sensing resistor* (R_S). Later, you will use the voltage across this resistor (V_{RS}) to calculate I_T and to observe its phase angle relative to V_S.

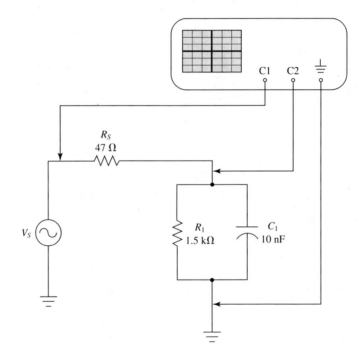

FIGURE 19.1

2. Set the function generator for a 10 V_{PP} sine wave at a frequency of 1 kHz. Measure the peak-to-peak voltage across the parallel network (V_{NET}), and record your measurement in Table 19.1. Measure the phase angle of the voltage across the parallel network relative to V_S, and record this value (θ_{NET}) in Table 19.1.

3. Repeat Step 2 for each frequency listed in Table 19.1. Make certain that V_S remains constant at 10 V_{PP} as you change the frequency.

TABLE 19.1

Frequency	V_{NET}	θ_{NET}	V_{RS}	θ_{RS}
1 kHz				
2 kHz				
5 kHz				
10 kHz				
20 kHz				
50 kHz				
100 kHz				

4. Reverse the components as shown in Figure 19.2. Repeat Steps 2 and 3 for the new circuit. Measure the magnitude of V_{RS} and its phase angle relative to V_S. Enter your results in Table 19.1.

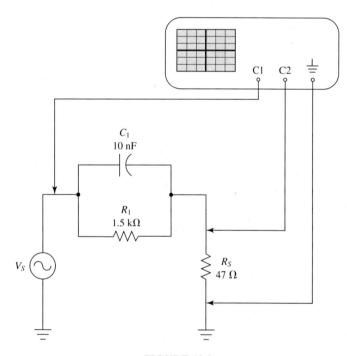

FIGURE 19.2

5. Calculate X_C for each frequency in Table 19.1, and record these values in Table 19.2.
6. Use V_{NET} and X_C to calculate the magnitude and phase angle of I_C at each frequency. Enter your results in Table 19.2.

TABLE 19.2

Frequency	X_C	I_C	I_{R1}	I_T	$I_T = \sqrt{I_C^2 + I_{R1}^2}$	$\theta = \tan^{-1}\left(\dfrac{-I_C}{I_{R1}}\right)$
1 kHz						
2 kHz						
5 kHz						
10 kHz						
20 kHz						
50 kHz						
100 kHz						

7. Use V_{NET} and R_1 to calculate the magnitude and phase angle of I_{R1} at each frequency. Enter your results in Table 19.2.
8. Use V_{RS} and R_S to calculate the magnitude and phase angle of I_T at each frequency. Enter your results in Table 19.2.
9. Use I_C and I_{R1} to calculate the magnitude and phase angle of I_T. Enter your results in Table 19.2.
10. Using your results from Table 19.2, draw the current phasor diagrams for this circuit at frequencies of 1 kHz, 10 kHz, and 100 kHz in Figure 19.3.

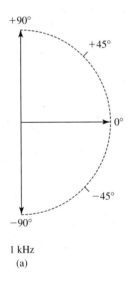

1 kHz

(a)

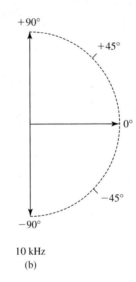

10 kHz

(b)

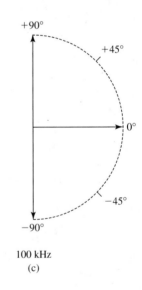

100 kHz

(c)

FIGURE 19.3

11. Use your results from Table 19.2 and Ohm's law to calculate the magnitude and phase angle of Z_T at each frequency shown in Table 19.3. Enter your results in the table.
12. Use your results from Table 19.2 to draw a curve representing the relationship between the current phase angle and frequency in Figure 19.4a.
13. Use your results from Table 19.3 to draw a curve representing the relationship between the impedance phase angle and frequency in Figure 19.4b.

TABLE 19.3

Frequency	$Z_T \angle \theta$
1 kHz	
2 kHz	
5 kHz	
10 kHz	
20 kHz	
50 kHz	
100 kHz	

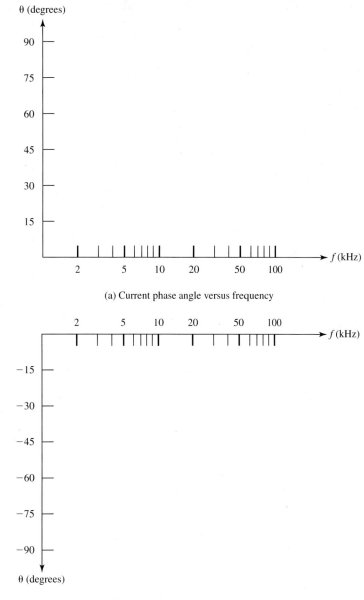

(a) Current phase angle versus frequency

(b) Impedance phase angle versus frequency

FIGURE 19.4

1. Refer to Table 19.1. As the frequency increased, did the total circuit current become more or less capacitive? Explain your answer.

2. Refer to Table 19.2. Explain how these results demonstrate that parallel *RC* circuits comply with Kirchhoff's current law.

3. Refer to the phasor diagrams that you drew in Step 10. Explain what happens to the two branch currents in a parallel *RC* circuit when the frequency changes. Be sure to comment on their respective phase angles.

4. Refer to the phase angle curves that you generated in Steps 12 and 13. Explain why the phase angle of current becomes more positive as frequency increases, while the phase angle for Z_T gets more negative.

Discussion

A simulator allows a circuit to be "tested" before construction in lab, or the results of a lab activity to be checked. It is sometimes beneficial to run the exercise with a different set of instruments to verify that the results are valid even though the instruments provide different numbers.

In practice, many ac current meters are limited in the results they can provide due to their limited frequency response and/or their internal resistance. An ac meter provides an *effective* (rms) value of voltage or current; the scope provides either peak or peak-to-peak values. Comparing the results from an *rms indicator* with those from a *peak* or a *peak-to-peak indicator* can provide both a cross-check of results and an opportunity to practice ac value conversions.

Procedure

1. Open file Ex19.1 from the Electronics Technology Fundamentals companion web site (www.prenhall.com/paynter). This is the same circuit as the one shown in Figure 19.1. Note that multimeters are being used as ac ammeters since they provide better resolution than current indicators.
2. Run the simulation for each frequency listed in Table 19.4, recording the values of I_T, I_R, and I_C in the table.
3. Using your simulation results, complete the last two columns of Table 19.4.

TABLE 19.4   Simulation Results and Calculations

Frequency	I_T	I_{R1}	I_C	$\sqrt{I_C^2 + I_{R1}^2}$	$\tan^{-1}\left(\dfrac{I_C}{I_R}\right)$
1 kHz					
2 kHz					
5 kHz					
10 kHz					
20 kHz					
50 kHz					
100 kHz					

Questions

1. Do the measured and calculated values of I_T closely match each other? Explain any discrepancies in the two values.

2. Can the currents listed in Table 19.4 be directly compared with those in Table 19.2? If not, what must be done to allow the comparison?

3. How do the simulator phase angles compare with those determined in the hardware lab procedure?

4. If the value of the current-sensing resistor shown in Figure 19.1 (which is in the position traditionally used to represent source resistance) is increased, what effect will this change have on the impedance and phase angle "seen" by the source?

Exercise 20

Series LC Circuits

OBJECTIVES

After completing this exercise, you should be able to:

- Determine the impedance of a series *LC* circuit.
- Demonstrate the characteristics of a series *LC* circuit when $X_L < X_C$.
- Demonstrate the characteristics of a series *LC* circuit when $X_L > X_C$.
- Demonstrate the characteristics of a series *LC* circuit when $X_L = X_C$.

DISCUSSION

As you have seen:

- Voltage *leads* current by 90° in a purely inductive circuit.
- Voltage *lags* current by 90° in a purely capacitive circuit.

As a result of these relationships, inductor voltage is 180° out of phase with capacitor voltage in a series *LC* circuit. This means that the applied voltage in a series *LC* circuit equals the *difference between V_L and V_C*. For Kirchhoff's voltage law to hold true, inductor and/or capacitor voltage must be *greater* in magnitude than the applied voltage.

There are three possible relationships between V_L and V_C. If $X_L > X_C$, the net series reactance (X_S) is *inductive*, and $V_L > V_C$. If $X_L < X_C$, the net series reactance is *capacitive*, and $V_C > V_L$. In the first part of this exercise, you will look at these two conditions.

The third relationship is a special case. If $X_L = X_C$, then $X_S = 0\ \Omega$ (assuming ideal components). Since V_L and V_C are 180° out of phase, the net voltage drop across the two reactive components is zero. In the second part of this exercise, you will take a preliminary look at this special-case condition known as *resonance*.

LAB PREPARATION

Review Sections 14.1 and 14.3 of *Electronics Technology Fundamentals*.

MATERIALS

1   dual-trace oscilloscope
1   function generator
1   protoboard
1   22 Ω resistor
1   22 nF capacitor
1   10 mH inductor

PROCEDURE

Part 1: Series LC Circuits

1. Construct the circuit shown in Figure 20.1. Channel 1 (C1) is connected to provide a display of V_S. Channel 2 (C2) is connected to provide a display of V_{RS}.

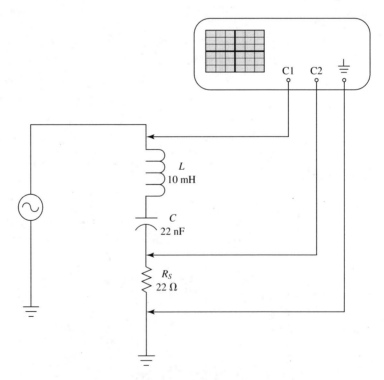

FIGURE 20.1

You will use the voltage across the sensing resistor (V_{RS}) to calculate I_T and measure its phase angle relative to V_S. Since $R_S \ll X_S$ at the various operating frequencies, we can treat this circuit as an *LC* circuit.

2. Set the function generator for a 10 V_{PP} sine wave at a frequency of 2 kHz. Measure the voltage across R_S and its phase angle relative to V_S. Record these values in Table 20.1. Calculate circuit current, and record this value in the table.

TABLE 20.1 Voltage and Current Values for the Circuit in Figure 20.1

Frequency	V_{RS}	θ	$I_T\angle\theta$
2 kHz			
5 kHz			
20 kHz			
50 kHz			

3. Repeat Step 2 for the remaining frequencies listed in Table 20.1. Make certain that V_S remains constant at 10 V_{PP} as you change the frequency.
4. Calculate the values of X_L and X_C for each of the four frequencies, and enter these values in Table 20.2. Then, solve for X_S for each frequency. Finally, use X_S and V_S to calculate I_T. (Since $R_S \ll X_S$ at these frequencies, you can ignore R_S in the I_T calculations.)

TABLE 20.2 Reactance and Current Values for the Circuit in Figure 20.1

Frequency	X_L	X_C	X_S	$I_T\angle\theta$
2 kHz				
5 kHz				
20 kHz				
50 kHz				

5. Use the values in Table 20.2 to calculate the values of V_L and V_C. Then, use these values to calculate V_S. Enter all these results in Table 20.3.

TABLE 20.3 Voltage Calculations for the Circuit in Figure 20.1

Frequency	V_L	V_C	V_S
2 kHz			
5 kHz			
20 kHz			
50 kHz			

6. Remove the inductor from the circuit, and measure its winding resistance. Record this value below. Return the inductor to the circuit. Then, calculate the resonant frequency for this circuit, and set the function generator to provide a 2 V_{PP} sine wave at this frequency.

$$R_W = \underline{\hspace{3cm}}$$

$$f_r = \frac{1}{2\pi\sqrt{LC}} = \underline{\hspace{3cm}}$$

7. Carefully adjust the frequency control of the signal generator until V_{RS} reaches its maximum value. Measure this frequency, and record it as the value of f_r in Table 20.4. Also, measure the peak-to-peak value of V_{RS} at this frequency and its phase angle relative to V_S. Record these values in Table 20.4 as well.

TABLE 20.4   Frequency Calculations for the Circuit in Figure 20.1

Frequency	V_{RS}	θ	$I_T\angle\theta$
$f_r - 1{,}000$ Hz =			
$f_r - 500$ Hz =			
$f_r - 200$ Hz =			
$f_r - 100$ Hz =			
$f_r =$			
$f_r + 100$ Hz =			
$f_r + 200$ Hz =			
$f_r + 500$ Hz =			
$f_r + 1{,}000$ Hz =			

8. Calculate the values of the remaining frequencies listed in Table 20.4.
9. For each frequency listed in Table 20.4, measure the peak-to-peak value of V_{RS} and its phase angle relative to V_S. Make sure that V_S remains constant at 2 V_{PP} as you vary the function generator output frequency.
10. Use V_{RS} and R_S to calculate the circuit current at each frequency listed in Table 20.4.

11. Use the values of I_T listed in Table 20.4 to plot a curve representing the relationship between I_T and operating frequency in Figure 20.2.

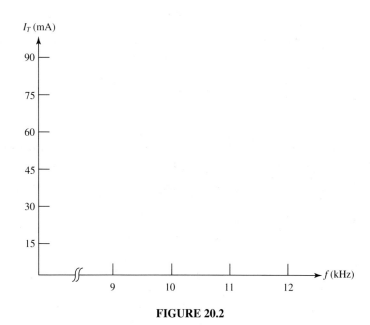

FIGURE 20.2

QUESTIONS & PROBLEMS

1. Refer to Table 20.2. Explain how and why θ changed when $X_L < X_C$ and when $X_L > X_C$.

2. Refer to Tables 20.1 and 20.2. Did both of your I_T calculations agree? If not, explain any discrepancies.

3. Explain how the series *LC* circuit conforms to Kirchhoff's voltage law using values from Table 20.3.

4. Why do you think the value of V_S was decreased from 10 V_{PP} to 2 V_{PP} in Part 2 of this exercise? (*Hint:* Think in terms of circuit power.) Support your answer with some calculations.

5. Refer to Table 20.4. How did θ change as frequency varied above and below f_r? Explain why these changes occurred.

6. Explain how the values of R_W and R_S affected the value of I_T *at resonance* for this circuit. (Feel free to use Ohm's law to prove your point.)

SIMULATION EXERCISE

Procedure

1. Open file Ex20.1 from the Electronics Technology Fundamentals companion web site (www.prenhall.com/paynter). This is the same circuit as the one shown in Figure 20.1.
2. Run the simulation. Measure the peak-to-peak value of V_{RS} and its phase angle relative to V_S, and enter these values in Table 20.5 (on page 164).
3. Use the simulator to insert the faults listed in Table 20.5 into the circuit. Run the simulation for each fault, and compare the performance of each faulted circuit with that of the normal circuit. Identify the major symptom of the failure and any additional symptoms that would assist you in determining the cause of this type of failure during later troubleshooting activities. For example, the meter readings and waveforms should be identified as *normal* or *abnormal* in such a way that you will recognize the conditions again when reviewing the chart in the future. Be sure to return each faulted component to its original state before inserting the next fault.

TABLE 20.5   Simulation Results

Condition	V_{RS}	θ	Primary Symptoms
Normal			None
L open			
C open			
R_S open			
L shorted			
C shorted			
R_S shorted			

Exercise 21

Parallel LC Circuits

OBJECTIVES

After completing this exercise, you should be able to:

- Determine the impedance of a parallel *LC* circuit.
- Demonstrate the characteristics of a parallel *LC* circuit when $X_L < X_C$.
- Demonstrate the characteristics of a parallel *LC* circuit when $X_L > X_C$.
- Demonstrate the characteristics of a parallel *LC* circuit when $X_L = X_C$.

DISCUSSION

In Exercise 20, you observed and analyzed series *LC* circuit operation. Since this was a series circuit, you focused on circuit voltages and Kirchhoff's voltage law. This exercise, however, deals with *parallel LC* circuits, so it should be no surprise that you will focus here on circuit currents and Kirchhoff's current law. (Isn't it amazing how the basic concepts of electronics, like Ohm's law and Kirchhoff's laws, keep cropping up?)

As with any *LC* circuit, three relationships are possible between X_L and X_C. The first part of this exercise will look at conditions when $X_L \neq X_C$. Even though you are primarily concerned with circuit currents, you will be measuring voltages using the oscilloscope and then *calculating* the resulting currents. For this reason, you will again employ a current-sensing resistor to determine the magnitude and phase of the total circuit current (I_T).

Since voltage is the same across the branches of any parallel circuit, the branch currents are 180° out of phase in a parallel *LC* circuit. As you know:

- Inductor current *lags* voltage by 90°.
- Capacitor current *leads* voltage by 90°.

Thus, the total current in a parallel *LC* circuit equals the *difference between* I_C and I_L. The phase of I_T (relative to V_S) depends on whether I_L or I_C is the dominant current, as you will see in this exercise.

When X_L and X_C are equal, you again have a special-case situation known as *resonance*. When a parallel *LC* circuit is operated at its resonant frequency, $I_C - I_L = 0$ A (assuming ideal components). In practice, of course, this is never the case. In the second part of this exercise, you will look briefly at parallel resonance.

LAB PREPARATION

Review Sections 14.2 and 14.3 of *Electronics Technology Fundamentals*.

MATERIALS

1 dual-trace oscilloscope
1 function generator
1 protoboard
1 10 Ω resistor
1 22 nF capacitor
1 10 mH inductor

PROCEDURE

Part 1: Parallel LC Circuits

1. Construct the circuit shown in Figure 21.1. Channel 1 (C1) is connected to display V_S. Channel 2 (C2) is connected to display V_{RS} (which is used to calculate I_T and to measure its phase angle relative to V_S).

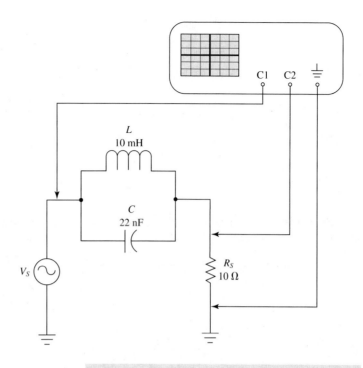

Note: This circuit is technically a series-parallel *RLC* circuit. The value of R_S, however, is so low that it does not significantly affect the characteristics of the parallel *LC* circuit.

FIGURE 21.1

2. Set the function generator for a 10 V_{PP} sine wave output at a frequency of 2 kHz. Measure the peak-to-peak value of V_{RS}, and its phase angle relative to V_S. Record these values in Table 21.1. Then, calculate and record the value of the circuit current.

TABLE 21.1   Measured Values for the Circuit Shown in Figure 21.1

Frequency	V_{RS}	θ	$I_T \angle \theta$
2 kHz			
5 kHz			
20 kHz			
50 kHz			

3. Repeat Step 2 for the remaining frequencies listed in Table 21.1. Make certain that V_S remains constant at 10 V_{PP} as you change the frequency.

4. Calculate X_L and X_C for each frequency listed in Table 21.1, and enter these values in Table 21.2. Use these values and V_S to calculate I_L and I_C.

TABLE 21.2 Reactance and Current Calculations for the Circuit Shown in Figure 21.1

Frequency	X_L	X_C	I_C	I_L	I_T
2 kHz					
5 kHz					
20 kHz					
50 kHz					

5. Using I_L and I_C, calculate and record the value of I_T at each frequency listed in Table 21.2. (Since $R_S \ll X_P$ at these frequencies, we can ignore V_{RS} in our current calculation.)

6. Use the values in Table 21.2 to calculate V_L and V_C. Enter your results in Table 21.3.

TABLE 21.3 Voltage Values for the Circuit Shown in Figure 21.1

Frequency	V_L	V_C
2 kHz		
5 kHz		
20 kHz		
50 kHz		

Part 2: Parallel Resonant LC Circuits

7. Calculate the resonant frequency for the circuit shown in Figure 21.1, and set the function generator for a 20 V_{PP} sine wave output at that frequency.

$$f_r = \frac{1}{2\pi\sqrt{LC}} = \underline{\hspace{3cm}}$$

8. Carefully adjust the frequency control of the signal generator until V_{RS} reaches its *minimum* value. Measure this frequency, and record it as the value of f_r in Table 21.4. Also, measure the peak-to-peak value of V_{RS} at this frequency and its phase angle relative to V_S. Record these values in Table 21.4 as well.

9. Calculate the remaining frequencies listed in Table 21.4.

10. For each frequency listed in Table 21.4, measure the peak-to-peak value of V_{RS} and its phase angle relative to V_S. Make sure that V_S remains constant at 20 V_{PP} as you vary the function generator output frequency. Use V_{RS} and R_S to calculate the circuit current at each frequency listed in Table 21.4.

TABLE 21.4 Voltage, Current, and Impedance Values for the Circuit Shown in Figure 21.1

Frequency	V_{RS}	θ	$I_T\angle\theta$	$Z_T\angle\theta$
$f_r - 1000$ Hz =				
$f_r - 500$ Hz =				
$f_r - 200$ Hz =				
$f_r - 100$ Hz =				
$f_r =$				
$f_r + 100$ Hz =				
$f_r + 200$ Hz =				
$f_r + 500$ Hz =				
$f_r + 1000$ Hz =				

11. Use the I_T values from Table 21.4 and V_S to calculate Z_T at each frequency. Record these values in Table 21.4.

12. Use the values of Z_T listed in Table 21.4 to plot a curve representing the relationship between Z_T and operating frequency in Figure 21.2.

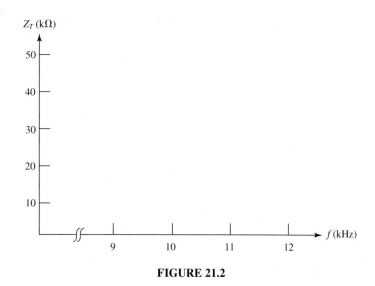

FIGURE 21.2

QUESTIONS & PROBLEMS

1. Refer to Table 21.1. Explain how and why the phase angle of I_T changes when $X_L < X_C$ and when $X_L > X_C$.

2. Explain how the parallel *LC* circuit conforms to Kirchhoff's current law using values from Table 21.2.

3. Refer to Table 21.4. Did I_T and Z_T come close to their theoretical values of 0 A and ∞ Ω, respectively? Why or why not?

4. Refer to Table 21.4. Describe the changes in the impedance and current phase angles that occur as the frequency varies above and below f_r. Explain why the two angles changed in opposite directions. (*Hint:* Think in terms of Ohm's law.)

SIMULATION EXERCISE

Procedure

1. Open file Ex21.1 from the Electronics Technology Fundamentals companion web site (www.prenhall.com/paynter). This is the same circuit as the one shown in Figure 21.1 with one difference. Since the inductor in the simulator is ideal, a 10Ω resistor has been added in series so that it acts more like a practical inductor.
2. Run the simulation at 2 kHz, and use the oscilloscope and ammeter to verify the circuit operation. Measure the peak-to-peak value of V_R and its phase angle relative to V_S. Record these normal values in Table 21.5.

TABLE 21.5   Simulation Results

Condition	V_{RS}	θ	Primary Symptoms
Normal			None
L open			
C open			
L shorted			
C shorted			

3. Use the simulator to insert the faults listed in Table 21.5 into the circuit. Run the simulation for each fault, and compare the performance of each faulted circuit with that of the normal circuit. Identify the major symptom of the failure and any additional symptoms that would assist you in determining the cause of this type of failure during later troubleshooting activities. (For example, the meter readings and waveforms should be identified as *normal* or *abnormal* in such a way that you will recognize the conditions again when reviewing the chart in the future.) Be sure to return each faulted component to its original state before inserting the next fault.

Exercise 22

Series RLC Circuits

OBJECTIVES

After completing this exercise, you should be able to:

- Determine the impedance of a series *RLC* circuit.
- Demonstrate the frequency response characteristics of a series *RLC* circuit.
- Calculate P_R, P_X, P_{APP}, and the power factor for a series *RLC* circuit.

DISCUSSION

You have already had some lab exposure to *RLC* circuits. The circuits used in Exercises 20 and 21 had inductors, capacitors, and resistors. The current-sensing resistors used in these circuits, however, had very low values compared to the circuit reactance values. As a result, they had little impact on the circuits. In this exercise, you will examine series *RLC* circuits in which the resistor comprises a significant part of the total circuit impedance.

The first part of this exercise focuses on the frequency response of a series *RLC* circuit. The frequency response of an *RLC* circuit is determined by the relationship between the net series reactance (X_S) and the total circuit resistance.

The second part of this exercise involves some power calculations. Unfortunately, you most likely do not have the equipment to measure power directly, so you will be using measured voltage and current values to calculate the various power values. You will also examine the effects of changes in frequency on apparent power, its phase angle, and the power factor of the circuit.

LAB PREPARATION

Review:

- Section 14.4 of *Electronics Technology Fundamentals*.
- Sections 11.2 and 13.2 of *Electronics Technology Fundamentals*. (These sections deal with power.)

MATERIALS

1 dual-trace oscilloscope
1 function generator
1 protoboard
1 560 Ω resistor
1 10 nF capacitor
1 1 mH inductor

PROCEDURE

Part 1: Series RLC Circuits

1. Construct the circuit shown in Figure 22.1. Channel 1 (C1) is connected to display V_S. Channel 2 (C2) is connected to display V_R.

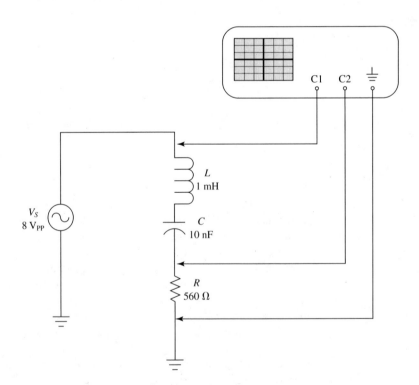

FIGURE 22.1

2. Calculate X_L, X_C, and X_S for the circuit at each frequency shown in Table 22.1. Enter your results in the table.

TABLE 22.1

Frequency	X_L	X_C	$X_S = X_L - X_C$
25 kHz			
50 kHz			
100 kHz			

3. Set the function generator for an 8 V_{PP} sine wave output at a frequency of 25 kHz. Measure the peak-to-peak value of V_R and its phase angle with respect to V_S, and record these values in Table 22.2.

TABLE 22.2 Voltage Measurements and Calculations for the Circuit Shown in Figure 22.1

Frequency	$V_R \angle \theta$	$V_L \angle \theta$	$V_C \angle \theta$	$V_X \angle \theta$	V_S	θ
25 kHz						
50 kHz						
100 kHz						

4. Repeat the magnitude and phase angle measurements for V_R at circuit frequencies of 50 kHz and 100 kHz, making certain that V_S stays constant at 8 V_{PP}.
5. Change the order of the components as shown in Figure 22.2. Then, repeat Steps 3 and 4, this time measuring the magnitude of V_L and its phase angle relative to V_S.

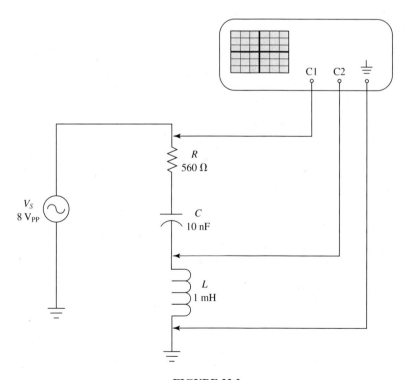

FIGURE 22.2

6. Change the order of the components as shown in Figure 22.3. Then, repeat Steps 3 and 4, this time measuring the magnitude of V_C and its phase angle relative to V_S.

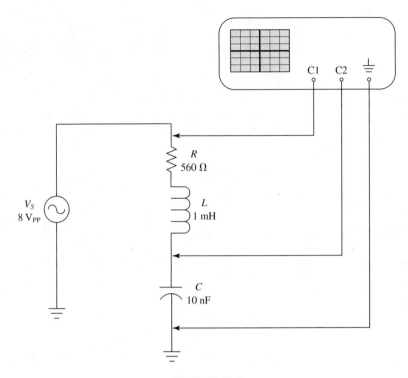

FIGURE 22.3

7. Use your reactive voltage measurements from Table 22.2 to calculate V_X at each frequency listed.
8. Now, use V_R and V_X to calculate V_S and its phase angle at each frequency listed.
9. Use your results from Table 22.2 to calculate I_T and its phase angle at each frequency listed. Make sure that you include the phase angles of V_R in your calculations. Enter your results in Table 22.3. Then, use I_T and the source voltage to calculate Z_T and its phase angle, and enter your results in the table.

TABLE 22.3   Calculated Current and Impedance Values

Frequency	$I_T \angle \theta$	$Z_T \angle \theta$	$Z_T = \sqrt{R^2 + X_S^2}$	$\theta = \tan^{-1}\dfrac{X_S}{R}$
25 kHz				
50 kHz				
100 kHz				

10. Now, use X_S and R to calculate Z_T and its phase angle for each frequency listed in Table 22.3, and enter your results in the table.

Part 2: Power in a Series RLC Circuit

11. Use your voltage, reactance, and resistance values to calculate the values of reactive and resistive power. Remember to convert your values to rms. Record these values in Table 22.4.

TABLE 22.4   Circuit Power Values

Frequency	P_R	P_L	P_C	P_X	$P_{APP} = \sqrt{P_R^2 + P_X^2}$	$\theta = \tan^{-1}\dfrac{P_X}{P_R}$
25 kHz						
50 kHz						
100 kHz						

12. Solve for P_X, and then solve for apparent power and its phase angle.
13. Use V_S and your values for I_T (from Table 22.3) to solve for P_{APP}. Remember to convert your values to rms. Be sure to use I_T as your 0° phase reference to get the correct phase angle for P_{APP}.

$$P_{APP} \text{ @ } 25 \text{ kHz} = \underline{\hspace{3cm}}$$

$$P_{APP} \text{ @ } 50 \text{ kHz} = \underline{\hspace{3cm}}$$

$$P_{APP} \text{ @ } 100 \text{ kHz} = \underline{\hspace{3cm}}$$

14. Use your results from Table 22.4 to solve for the power factor (PF) for each frequency.

$$PF \text{ @ } 25 \text{ kHz} = \underline{\hspace{3cm}}$$

$$PF \text{ @ } 50 \text{ kHz} = \underline{\hspace{3cm}}$$

$$PF \text{ @ } 100 \text{ kHz} = \underline{\hspace{3cm}}$$

QUESTIONS & PROBLEMS

1. Refer to your results in Table 22.2. Which reactance was the dominant value at 25 kHz? Why? Which reactance was the dominant value at 100 kHz? Why?

2. Refer to your results in Table 22.2. At which frequency did the circuit act mostly resistive? Explain why.

3. Refer to your results in Table 22.3. What happened to the phase angle of Z_T when the frequency increased? Explain why.

4. If the value of resistance in this circuit were to increase, what would happen to the phase angle of I_T?

5. Refer to your P_{APP} calculations in Table 22.4 and Step 13. Do they agree? If not, explain why you feel there is a discrepancy.

6. Refer to your results from Step 14. Explain why the power factor at 25 kHz was very similar to that at 100 kHz.

SIMULATION EXERCISE

Discussion

The hardware portion of this exercise requires a lot of rewiring to keep the component of interest referenced to ground for the oscilloscope measurements. The simulator provides a useful component that allows differential measurements and solves the grounding problem inherent in the hardware exercise. The component is a *voltage-controlled voltage source* (VCVS). In essence, the VCVS is an ideal voltage source that provides an output that is proportional (in both magnitude and phase) to a floating input voltage. The *voltage ratio* of the VCVS is a setting that can be used to increase or decrease the output voltage that is produced for a given input voltage. The common practice in this application is to use the default ratio of 1 V/V, which results in an output voltage that is equal to the floating input voltage. As you will see, the VCVS allows you to focus more on the measurements and less on rewiring the circuit.

Procedure

1. Open file Ex22.1 from the Electronics Technology Fundamentals companion web site (www.prenhall.com/paynter). This is the same circuit as the one shown in Figure 22.1, but with two voltage-controlled voltage sources connected to allow you to measure V_S, V_L, V_C, and V_R without rewiring. Note that Channel A is measuring V_R. This allows you to use I_T as the phase reference since V_R and I_T are in phase.
2. Run the simulation and determine the phase angle of V_S with respect to I_T (V_R). Record this value in Table 22.5. Determine if V_S is leading or lagging I_T.

TABLE 22.5   Simulation Results

Frequency	θ (V_S)	θ (V_C)	θ (V_L)	Is V_S Leading or Lagging I_T?
25 kHz				
50 kHz				
100 kHz				

3. Change the source frequency, first to 50 kHz, then 100 kHz, and repeat Step 2. Record your measurements in Table 22.5.
4. Move the lead for Channel B to measure V_L and repeat Step 3 for all three frequencies. Record your measurements in Table 22.5.

5. Move the lead for Channel B to measure V_C and repeat Step 3 for all three frequencies. Record your measurements in Table 22.5.

6. Open file Ex22.2 from the Electronics Technology Fundamentals companion web site (www.prenhall.com/paynter). This is the same circuit as in File Ex22.1, but a four-input oscilloscope is being used. This allows you to measure V_S, V_L, V_C, and V_R simultaneously. This also allows you to see the relative phase angles of all four waveforms.

7. Run the simulation at the three frequencies listed in Table 22.5 and substantiate that the magnitude and relative phase measurements recorded in the table are correct. Note that V_C and V_L are always 180° out of phase, regardless of the frequency.

Questions

1. When $X_C > X_L$ in a series *RLC* circuit, the net series reactance (X_S) appears capacitive. When $X_C < X_L$, X_S appears inductive. Explain how the simulation exercise demonstrates these relationships.

2. Between which two frequencies did the phase angle reverse (from lead to lag, or vice versa)? Support your answer with reference to specific measurements.

3. Using values from Table 22.5, demonstrate the phase relationship between V_L and V_C.

Exercise 23

Parallel RLC Circuits

OBJECTIVES

After completing this exercise, you should be able to:

- Analyze a parallel *RLC* circuit.
- Demonstrate the frequency response characteristics of a parallel *RLC* circuit.
- Calculate P_R, P_X, P_{APP}, and the power factor for a parallel *RLC* circuit.

DISCUSSION

By now, you should be familiar with the differences between series and parallel circuits. Even though more components have been added to the mix, things really haven't changed that much. Parallel *RLC* circuits comply with all the basic operating rules of other parallel circuits. Voltage is equal across the various branches, and Kirchhoff's current law is as valid as ever. The only thing to keep in mind is the relative phase angles of the currents, which must be considered in your circuit calculations.

In the first part of this exercise, you will look at the frequency response characteristics of a parallel *RLC* circuit. In the second part of the exercise, you will look briefly at power in parallel *RLC* circuits.

One final note: As you have done many times before, you will be using a current-sensing resistor (R_S). Because this resistor has a very low value, it can be ignored in your calculations. As before, it is used simply to measure the total circuit current and its phase angle relative to V_S.

LAB PREPARATION

Review:

- Section 14.4 of *Electronics Technology Fundamentals*.
- Sections 11.2 and 13.2 of *Electronics Technology Fundamentals*. (These sections deal with power.)

MATERIALS

- 1 dual-trace oscilloscope
- 1 function generator
- 1 protoboard
- 2 resistors: 560 Ω and 10 Ω
- 1 10 nF capacitor
- 1 1 mH inductor

PROCEDURE

Part 1: Parallel RLC Circuits

1. Calculate the values of X_L, X_C, and X_P for the frequencies listed in Table 23.1, and enter your results in the table.

TABLE 23.1

Frequency	X_L	X_C	$X_P = \dfrac{X_L X_C}{X_L + X_C}$
25 kHz			
50 kHz			
100 kHz			

2. Construct the circuit shown in Figure 23.1. Channel 1 (C1) is connected to display V_S. Channel 2 (C2) is connected to display V_{RS}.
3. Set the function generator for an 8 V_{PP} sine wave at a frequency of 25 kHz. Measure the peak-to-peak value of V_{RS} and its phase angle with respect to V_S. Record these values in Table 23.2.

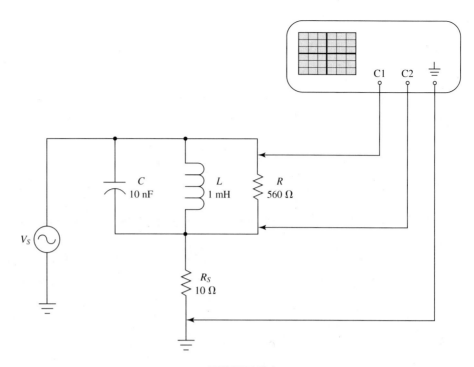

FIGURE 23.1

TABLE 23.2   Voltage and Current Values

Frequency	$V_{RS} \angle \theta$	$I_T \angle \theta$
25 kHz		
50 kHz		
100 kHz		

4. Repeat Step 3 for frequencies of 50 kHz and 100 kHz. Make sure that V_S stays constant at 8 V_{PP} as you change the operating frequency.
5. Use the values of V_{RS} and R_S to calculate I_T at each frequency listed in Table 23.2.
6. Change the order of the components as shown in Figure 23.2. Repeat Steps 3 and 4, measuring the peak-to-peak value of V_{RLC} and its phase angle relative to V_S. Enter these values in Table 23.3.

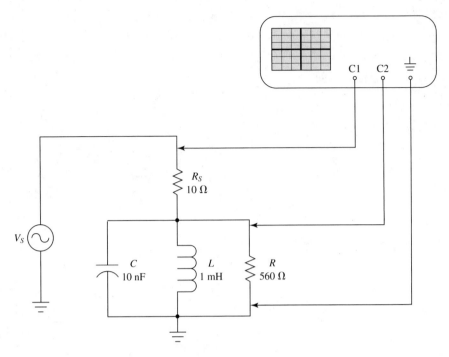

FIGURE 23.2

TABLE 23.3   Voltage and Current Values for the Circuit Shown in Figure 23.2

Frequency	$V_{RLC} \angle \theta$	I_R	I_L	I_C
25 kHz				
50 kHz				
100 kHz				

7. Using V_{RLC} and the component reactance (or resistance) values, calculate the values of I_R, I_L, and I_C for each frequency listed in Table 23.3.
8. Use the values of I_L and I_C to calculate I_{LC}, and enter these values in Table 23.4. Then, calculate I_T and its phase angle from the values of I_R and I_{LC}. Record these values in Table 23.4 as well.

TABLE 23.4   Current Calculations for the Circuit Shown in Figure 23.2

Frequency	I_{LC}	$I_T = \sqrt{I_R^2 + I_{LC}^2}$	$\theta = \tan^{-1}\dfrac{I_{LC}}{I_R}$
25 kHz			
50 kHz			
100 kHz			

9. Use the values of X_P from Table 23.1 and R to calculate Z_P for each frequency listed in Table 23.5. Then, use Z_P and V_S to calculate I_T. (*Note:* Because $R_S \ll Z_P$ you can ignore R_S in your I_T calculations.) Record your results in Table 23.5.

TABLE 23.5 Circuit Impedance and Current Values

Frequency	$Z_P = \dfrac{R \times X_P}{\sqrt{R^2 + X_P^2}}$	$\theta = \tan^{-1} \dfrac{R}{X_P}$	$I_T \angle \theta$
25 kHz			
50 kHz			
100 kHz			

Part 2: Power in Parallel RLC Circuits

10. Use the values in Table 23.3 to calculate component power. Record your results in Table 23.6.

TABLE 23.6 Power Values for the Parallel *RLC* Circuit

Frequency	P_R	P_L	P_C	P_X	$P_{APP} = \sqrt{P_R^2 + P_X^2}$	$\theta = \tan^{-1} \dfrac{P_X}{P_R}$
25 kHz						
50 kHz						
100 kHz						

11. Solve for P_X, P_{APP}, and the phase angle of P_{APP}.
12. Use V_S and your values for I_T from Table 23.2 to solve for P_{APP}. Remember to convert your values to rms. Be sure to include the phase angle of I_T.

$$P_{APP} @ 25 \text{ kHz} = \text{_____}$$

$$P_{APP} @ 50 \text{ kHz} = \text{_____}$$

$$P_{APP} @ 100 \text{ kHz} = \text{_____}$$

13. Use your results from Table 23.6 to solve for the power factor (PF) at each of the frequencies listed.

$$PF @ 25 \text{ kHz} = \text{_____}$$

$$PF @ 50 \text{ kHz} = \text{_____}$$

$$PF @ 100 \text{ kHz} = \text{_____}$$

QUESTIONS & PROBLEMS

1. Refer to Table 23.2. At each frequency listed, state whether the circuit is primarily inductive, capacitive, or resistive. Explain your reasoning.

2. Refer to Table 23.4. Explain how your results substantiate (or refute) Kirchhoff's current law.

3. Refer to your I_T calculations in Tables 23.4 and 23.5. Do the two current calculations agree? If not, explain any variations.

4. Refer to Table 23.6. Explain why you can *add* power values in a parallel circuit to solve for P_{APP} but cannot use this same approach to solve for impedance.

5. Refer to your results from Step 13. At which frequency was the power factor the highest? Explain why.

Discussion

Oscilloscopes and voltmeters only allow you to measure the voltage across a given component. However, if you connect a resistor in series with that component, you can use Ohm's law, and the voltage across the resistor, to calculate the current through the given component. Resistors that are used to determine the current through a given component are called *metering resistors* or *current-sensing resistors*. Installing a sensing resistor should never substantially change the performance of the circuit under test. To ensure that this does not occur, the sensing resistance should always be less than 10% of the total impedance of any series component(s).

This simulation has a current-sensing resistor added to three of the four branches of a parallel *RLC* circuit. A VCVS has been added to the I_T current-sensing resistor to provide a phase correction.

Procedure

1. Open file Ex23.1 from the Electronics Technology Fundamentals companion web site (www.prenhall.com/paynter). This is the same circuit as the one shown in Figure 23.1, but with the changes listed in the discussion.
2. Run the simulation and determine the phase angle of I_T (V_{RS}) with respect to V_S. Record this value in Table 23.7. Determine if I_T is leading or lagging V_S.

TABLE 23.7   Simulation Results

Frequency	θ (I_T)	θ (I_C)	θ (I_L)	θ (I_R)	I_T Leading or Lagging V_S?
25 kHz					
50 kHz					
100 kHz					

3. Change the source frequency, first to 50 kHz, then 100 kHz, and repeat Step 2. Record your measurements in Table 23.7.
4. Move the lead for Channel B to measure I_C (V_{RC}) and repeat Step 3 for all three frequencies. Record your measurements in Table 23.7.
5. Move the lead for Channel B to measure I_L (V_{RL}) and repeat Step 3 for all three frequencies. Record your measurements in Table 23.7.
6. Move the lead for Channel B to measure I_R (V_{Rl}) and repeat Steps 2 and 3 for all three frequencies. Record your measurements in Table 23.7.

Questions

1. When $X_C < X_L$ in a parallel *RLC* circuit, the net parallel reactance (X_P) appears capacitive. When $X_C > X_L$, X_P appears inductive. Explain how the simulation exercise demonstrates these relationships.

2. Between which two frequencies did the phase angle of I_T reverse (from lead to lag, or vice versa)? Support your answer with reference to specific measurements.

3. Using values from Table 23.7, demonstrate the phase relationship between I_L and I_C.

4. In this simulation, the VCVS was used to correct a phase problem inherent in displaying I_T. Specifically, the VCVS provided a 180° phase shift to the I_T signal. Explain why this polarity reversal was necessary. (*Hint:* Trace the current paths, and compare the polarity of the V_{R_S} voltage with other voltage measurements.

5. No current-sensing resistor was added to the 560 Ω branch. Explain why it was not needed in this branch.

Exercise 24

RC and RL Low-Pass Filters

After completing this exercise, you should be able to:

- Determine the cutoff frequency (f_C) of an *RC* or *RL* low-pass filter.
- Draw the frequency response curve for an *RC* or *RL* low-pass filter.
- Determine the roll-off rate for an *RC* or *RL* low-pass filter.
- Demonstrate the effect that a change in load has on the gain and frequency response of an *RC* or *RL* low-pass filter.

DISCUSSION

There are many types of filters, each with its own characteristics. Even limiting your focus to passive *RC, RL,* and *LC* filters, you could spend months investigating them. In the next three exercises, you will look at some of the most basic passive filters.

A *low-pass filter* is designed to pass all frequencies below a specific *cutoff frequency* (f_C). The cutoff frequency for a low-pass filter is the frequency at which load power is reduced to 50% of its midband (maximum) value. It also can be defined as the frequency at which load voltage is reduced to 70.7% of its midband value.

In the first part of this exercise, you will examine the frequency response and gain characteristics of an *RC* low-pass filter. As you know, the value of the load can have a significant impact on how its source circuit performs. In the second part of this exercise, you will look at a basic *RL* low-pass filter and how a change in load affects its cutoff frequency.

LAB PREPARATION

Review:

- Sections 15.1 through 15.4 of *Electronics Technology Fundamentals*.
- Appendix C on measuring the cutoff frequency of a circuit.

MATERIALS

1   dual-trace oscilloscope
1   function generator
1   protoboard
2   resistors: 1 kΩ and 10 kΩ
1   10 kΩ potentiometer
1   2.2 nF capacitor
1   10 mH inductor

PROCEDURE

Part 1: Low-pass RC Circuits

1. Construct the circuit shown in Figure 24.1. Channel 1 (C1) is connected to display V_S. Channel 2 (C2) is connected to display the load voltage (V_{RL}). Set the load initially to 10 kΩ.

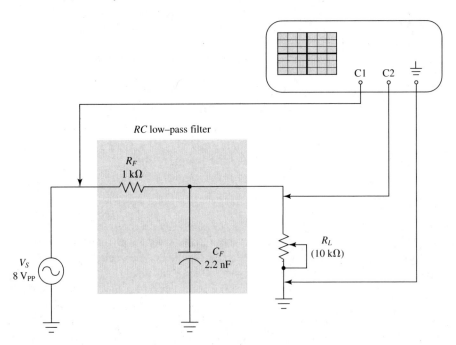

FIGURE 24.1   An *RC* low-pass filter.

2. Calculate the values below.

$$R_{TH} = R_F \parallel R_L = \underline{\hspace{3cm}}$$

$$f_C = \frac{1}{2\pi R_{TH}C} = \underline{\hspace{3cm}}$$

3. Set the function generator to produce an 8 V_{PP} sine wave output.
4. Use the technique introduced in Appendix C to measure the cutoff frequency of the circuit. Record this value in Table 24.1 as the cutoff frequency (f_C). Also, measure and record the peak-to-peak load voltage (V_{RL}) at f_C.

TABLE 24.1 Voltage Measurements and Calculations for the *RC* Low-Pass Filter

Frequency	V_{RL}	A_v
10 kHz		
20 kHz		
50 kHz		
$f_C =$		
100 kHz		
200 kHz		
500 kHz		
1 MHz		

5. Measure V_{RL} at each frequency listed in Table 24.1. Make certain that V_S remains at 8 V_{PP} as you change frequencies. Record your measurements in Table 24.1.
6. Calculate the value of A_v for the circuit at each frequency listed in Table 24.1, and record your results in the table.
7. Use your results in Table 24.1 to plot the frequency response curve for this circuit in Figure 24.2.

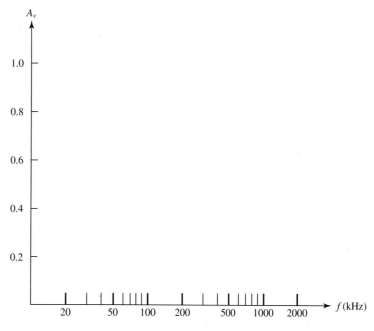

FIGURE 24.2

8. Construct the circuit shown in Figure 24.3. The load should be set initially to 10 kΩ.

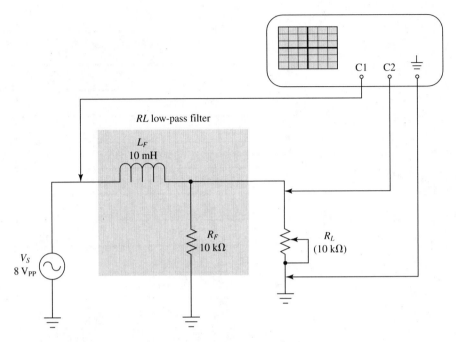

FIGURE 24.3   An *RL* low-pass filter.

9. Calculate and record the values below.

$$R_{TH} = R_F \parallel R_L = \underline{\hspace{3cm}}$$

$$f_C = \frac{R_{TH}}{2\pi L} = \underline{\hspace{3cm}}$$

10. Use the technique introduced in Appendix C to measure the cutoff frequency of the circuit. Record this value in Table 24.2 as the cutoff frequency (f_C).

TABLE 24.2   Cutoff Frequencies for the *RL* Low-Pass Filter

R_L	f_C
10 kΩ	
5 kΩ	
2 kΩ	
1 kΩ	
500 Ω	

11. Repeat Step 10 for each load value listed in Table 24.2, and record your measurements in the table.

QUESTIONS & PROBLEMS

1. Compare the calculated value of f_C from Step 2 with the measured value in Table 24.1. Do these two values agree? If not, explain any discrepancy between them.

2. Refer to Table 24.1 and Figure 24.2. Convert the A_v values at 100 kHz and 1 MHz to decibel (dB) form. Then, calculate the roll-off rate of the filter for this decade.

$$A_{v(dB)} \ @ \ 100 \ kHz = \underline{\hspace{3cm}}$$

$$A_{v(dB)} \ @ \ 1 \ MHz = \underline{\hspace{3cm}}$$

$$\Delta A_{v(dB)} = \underline{\hspace{3cm}}$$

Is your calculated value of $\Delta A_{v(dB)}$ consistent with the expected value of 20 dB/decade? If not, explain any discrepancy.

3. Refer to Table 24.2. Explain how the cutoff frequency of a filter is affected by load resistance, citing specific examples from the table.

SIMULATION EXERCISE

Discussion

One of the best instruments for simulating the frequency response of circuits is the Bode plotter. This instrument provides a visual representation of changes in gain per unit change in frequency. The vertical axis in the display represents the dB voltage gain, given as

$$A_{v(\text{dB})} = 20 \log \frac{V_{\text{out}}}{V_{\text{in}}}$$

Note that the Bode plotter displays calculated values, whereas a true Bode plot is flat until the critical frequency is reached and then rolls off at a specific rate or slope.

Before beginning this simulation, several terms must be defined:

- **3 dB frequency:**   The frequency that produces a response level that is 3 dB below the reference level.
- **Reference level:**   The nearly constant gain that a circuit produces over a wide range of frequencies. (This is also known as *midband gain*.)

Procedure

1. Open file Ex24.1 from the Electronics Technology Fundamentals companion web site (www.prenhall.com/paynter). This is the same circuit as the one shown in Figure 24.1, but with one major difference. You will use a Bode plotter instead of an oscilloscope to make your measurements.
2. Open the Bode plotter and run the simulation. Note that despite the frequency setting of the voltage source, the circuit has been swept from 5 kHz to 200 kHz. This represents the horizontal scale of the Bode plot. The vertical scale is from 0 dB to −10 dB. Note that both axes are log scales.
3. Set the Bode plot cursor on the flat part of the curve, and determine the mid-band gain or reference level of this circuit.

    Reference level = _____ dB

4. Now, use the cursor to determine the cutoff frequency of this filter.

    $f_C =$ _____

    > *Note:* f_C is the frequency that is 3 dB below the reference level. For example, if the reference level is −0.84 dB, then f_C is the frequency at which the response curve has rolled off to
    >
    > $$A_{v(\text{dB})} = -0.84 \text{ dB} - 3 \text{ dB} = -3.84 \text{ dB}$$
    >
    > The cursor should be placed at the frequency closest to this level.

5. Set the Bode plotter mode to *phase* instead of magnitude. Note that the vertical scale automatically switches to linear. Move the cursor back to the cutoff frequency that you determined in Step 4, and record the phase angle at this frequency.

    $\theta =$ _____

6. Change the settings of the Bode plotter as follows:
   - Choose magnitude.
   - Vertical log: F = 0 dB; I = − 45 dB.
   - Horizontal log: F = 15 MHz; I = 1 kHz.
7. Run the simulation. Move your cursor to approximately $10f_C$, and record the dB gain at this frequency. Then, move the cursor to approximately $100f_C$, and record the gain at this frequency. Use these values to calculate the roll-off rate for the filter.

$$A_{v(\text{dB})} @ 10f_C = \underline{\hspace{3cm}}$$

$$A_{v(\text{dB})} @ 100f_C = \underline{\hspace{3cm}}$$

$$\text{Roll-off rate} = \underline{\hspace{3cm}} \text{ dB/decade}$$

8. Now, change the mode back to phase. Use the cursor to determine the phase angle one decade below f_C, at f_C, and one decade above f_C.

$$\theta = \underline{\hspace{3cm}} @ \frac{f_C}{10}$$

$$\theta = \underline{\hspace{3cm}} @ f_C$$

$$\theta = \underline{\hspace{3cm}} @ 10f_C$$

Questions

1. Calculate the theoretical midband gain of this circuit, and compare this value to your results from Step 3. Comment on any discrepancies.

2. Compare your results from Step 4 with the cutoff frequency recorded in Table 24.1. How would you explain any difference between the two frequencies?

3. Refer to your results in Step 8. Was the roll-off rate consistent with theoretical values? If not, explain why.

4. Refer to Step 7. Why did we not use frequencies of f_C and $10f_C$ to determine the roll-off rate? (*Hint:* Look at the slope of the curve at f_C.)

5. Explain, in your own words, why the phase angle was near 0° one decade below f_C and near −90° one decade above. (*Hint:* Think in terms of what effect the capacitor has on the circuit as the frequency increases.)

Exercise 25

RC and RL High-Pass Filters

After completing this exercise, you should be able to:

- Determine the cutoff frequency (f_C) of an *RC* or *RL* high-pass filter.
- Draw the frequency response curve for an *RC* or *RL* high-pass filter.
- Determine the roll-off rate for an *RC* or *RL* high-pass filter.
- Demonstrate the effect that a change in load has on an *RC* or *RL* high-pass filter.

DISCUSSION

High-pass filters are designed to pass all frequencies above a specified cutoff frequency (f_C). The only difference in circuit construction between high-pass and low-pass filters is the relative position of the filter resistor and the reactive component. Even the equation used to calculate the cutoff frequency for the two filters is the same. These filters also exhibit the same slope (20 dB/decade, or 6 dB/octave) when operated outside the pass band. In this exercise, you will examine these concepts and look at the relationship between phase angle and cutoff frequency.

LAB PREPARATION

Review:

- Section 15.5 of *Electronics Technology Fundamentals*.
- Appendix C on measuring the cutoff frequency of a circuit.

MATERIALS

1   dual-trace oscilloscope

1   function generator

1   protoboard

2   resistors: 2.7 kΩ and 10 kΩ

1   10 kΩ potentiometer

1   1 nF capacitor

1   10 mH inductor

PROCEDURE

Part 1: High-Pass RC Circuits

1.   Construct the circuit shown in Figure 25.1. Channel 1 (C1) is connected to display V_S. Channel 2 (C2) is connected to display the load voltage (V_{RL}). Set the potentiometer initially to 10 kΩ.

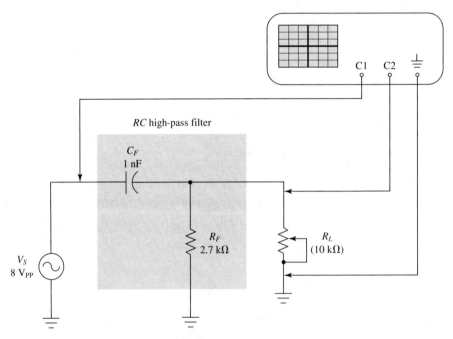

FIGURE 25.1

2.   Calculate the values below.

$$R_{TH} = R_F \parallel R_L = \underline{\hspace{3cm}}$$

$$f_C = \frac{1}{2\pi R_{TH}C} = \underline{\hspace{3cm}}$$

3.   Set the function generator for an 8 V$_{PP}$ sine wave output.

4.   Measure the values below, and record them in Table 25.1.

    • The cutoff frequency (f_C).

    • The phase angle between V_S and V_{RL} at f_C.

TABLE 25.1   *RC* High-Pass Filter Voltages and Gains

Frequency	V_{RL}	θ	A_v
10 kHz			
20 kHz			
50 kHz			
$f_C =$			
100 kHz			
200 kHz			
500 kHz			
1 MHz			

5. Repeat Step 4 for each frequency listed in Table 25.1. Make certain that V_S remains constant at 8 V_{PP} as you change the frequency. Record your measurements in Table 25.1.
6. Calculate A_v for each frequency, and record your results in Table 25.1.
7. Use your results from Table 25.1 to plot the frequency response curve for this circuit in Figure 25.2a.

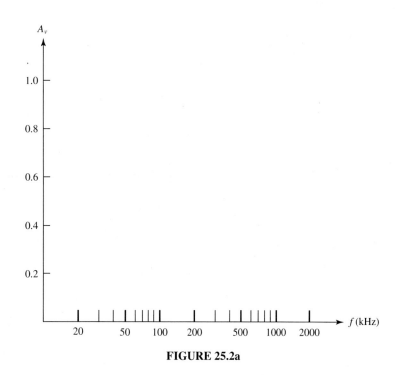

FIGURE 25.2a

8. Use your results from Table 25.1 to plot the phase angle versus frequency curve for this circuit in Figure 25.2b.

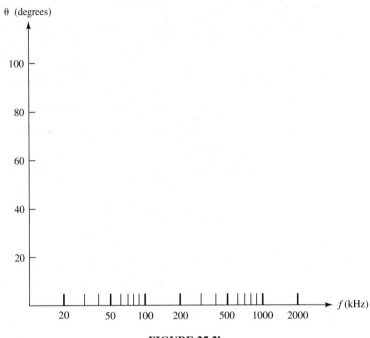

FIGURE 25.2b

Part 2: High-Pass RL Circuits

9. Construct the circuit shown in Figure 25.3.

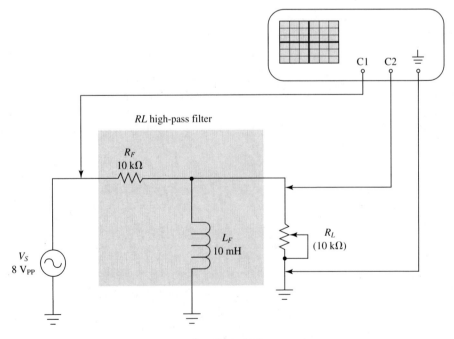

FIGURE 25.3

10. Calculate the values below.

$$R_{TH} = R_F \parallel R_L = \underline{\hspace{3cm}}$$

$$f_C = \frac{R_{TH}}{2\pi L} = \underline{\hspace{3cm}}$$

11. Measure the values below, and record them in Table 25.2.
    - The cutoff frequency (f_C).
    - The phase angle of V_{RL} relative to V_S.

TABLE 25.2

R_L	f_C	θ
10 kΩ		
5 kΩ		
2 kΩ		
1 kΩ		
500 Ω		

12. Repeat Step 11 for each load value listed in Table 25.2.

QUESTIONS & PROBLEMS

1. Refer to the calculated value of f_C from Step 2 and your measured value from Table 25.1. Do these two values agree? If not, explain any discrepancies.

2. Refer to Table 25.1 and Figure 25.2a. Convert the A_v values at 20 kHz and 10 kHz to dB form. Then, calculate the roll-off rate of the filter for this decade.

$$A_{v(\text{dB})} @ 20 \text{ kHz} = \underline{\hspace{3cm}} \text{ dB}$$

$$A_{v(\text{dB})} @ 10 \text{ kHz} = \underline{\hspace{3cm}} \text{ dB}$$

$$\Delta A_{v(\text{dB})} = \underline{\hspace{3cm}} \text{ dB/octave}$$

How does this roll-off rate compare to the *dB/decade* rate calculated for the low-pass filter in Exercise 24 (Question 2, page 193)?

3. Refer to Table 25.1. Describe the changes in phase angle above and below the cutoff frequency.

4. Refer to Tables 25.1 and 25.2. What was the phase angle of load voltage at f_C? Was the phase angle affected by changes in load resistance?

SIMULATION EXERCISE

Discussion

In this part of the exercise you will focus on one important difference between *RL* and *RC* high-pass filters. Although they have much in common, there is one major difference between these two circuits. When a filter is "*inserted*" into an existing circuit, it results in some "*insertion loss*". By this we mean that even in the passband of the filter, the output is lower than it would be if the filter was not there. The voltage in the passband is referred to as the reference voltage (V_{Ref}), and when compared to the input voltage of the filter, we can determine the reference level of the circuit, usually expressed in dB.

As you saw in the simulation portion of Exercise 24, one of the best instruments for simulating the frequency response characteristics of a circuit is the Bode plotter. Remember that the Bode plotter used in the simulator displays calculated values. A true Bode plot is flat until the critical frequency (f_C) is reached, and then rolls off at a specified rate.

Procedure

1. Open file Ex25.1 from the Electronics Technology Fundamentals companion web site (www.prenhall.com/paynter). These are the same circuits as those shown in Figures 25.1 and 25.3. The only difference is that you will use the Bode plotter, instead of an oscilloscope, to make your measurements.
2. Note that the settings on the Bode plotters for each circuit are slightly different. Both circuits are swept from 20 kHz to 2.5 MHz, but the vertical scale for the *RL* filter is set to 0 to −20 dB, while the vertical scale for the *RC* circuit is 0 to −10 dB.
3. Open both Bode plotters and run the simulation until you get a trace on both plotters. Move the cursor on both plotters to the far right side of the display. This is well into the pass-band of both filters and thus, represents the reference level for the filters. Record the reference level for both circuits below.

Reference level = _____ dB (*RC* filter)

Reference level = _____ dB (*RL* filter)

4. Determine the cutoff frequencies for each circuit, and record these values. (*Remember:* The cutoff frequency is the frequency at which the output is 3 dB below the reference level.)

f_C = _____ kHz (*RC* filter)

f_C = _____ kHz (*RL* filter)

5. Make the following changes to the *RL* filter: $L_F = 1$ mH, and $R_F = 510\ \Omega$. Run the simulation and measure the following values for the *RL* circuit:

V_{Ref} = _____ dB (*RL* filter)

f_C = _____ kHz (*RL* filter)

6. Call up both Bode plotter displays, and select "phase" for both plotters. How do the phase characteristics of the two filters compare?

7. Return the *RL* filter components to their original values ($R_F = 10$ kΩ, $L_F = 10$ mH), and run the simulation again. Repeat Step 6. Record your phase observations below.

1. Refer to your results from Steps 3 and 4. Explain why there was such a large difference in V_{Ref} between the *RL* and *RC* filters even though the cutoff frequencies were so close.

2. Refer to your results from Steps 4 and 5. Explain why there was such a large difference in V_{Ref} between these two *RL* filters even though the cutoff frequencies were so close.

3. Which of the two *RL* filters is the best design? Support your answer.

4. Refer to your answers to Steps 6 and 7. Explain why you think the phase characteristics of the circuits were so similar.

Exercise 26

LC Bandpass Filters

OBJECTIVES

After completing this exercise, you should be able to:

- Measure the resonant frequency (f_r) of an *LC* bandpass filter.
- Measure the bandwidth (BW) of an *LC* bandpass filter.
- Measure the effect of source impedance on the BW and Q of an *LC* bandpass filter.
- Measure the effect of load resistance on the BW and Q of an *LC* bandpass filter.

DISCUSSION

Both *bandpass* and *band-stop* (notch) filters use the resonant characteristics of series and parallel *LC* circuits to either pass or block a band of frequencies. A series resonant *LC* circuit has very low impedance at its resonant frequency, but a parallel *LC* circuit has very high impedance at resonance. The type of resonant circuit, and its relative position to the load, determine the type of filter (bandpass or notch). This is illustrated in Figure 26.1.

Regardless of the type of filter involved, the principle is the same. Each circuit uses the resonant characteristics of the *LC* circuit to couple, block, or short out an input signal. The series bandpass filter couples the input signal to the load, whereas the shunt bandpass filter shorts out all signals outside of its own resonant passband. The series notch filter blocks all frequencies within its passband while coupling all other frequencies to the load. The shunt notch filter shorts out only those frequencies within its passband.

Because the operation of bandpass and notch filters work on the same principles, you will focus on the bandpass filter.

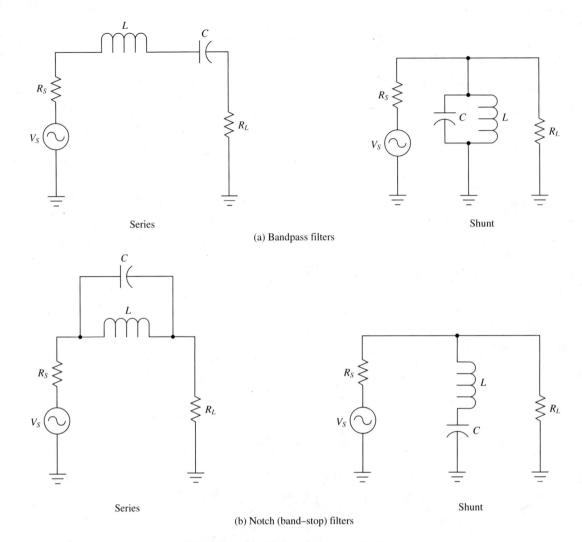

Series

Shunt

(a) Bandpass filters

Series

Shunt

(b) Notch (band–stop) filters

FIGURE 26.1   *LC* bandpass and notch filters.

LAB PREPARATION

Review:

- Section 15.6 of *Electronics Technology Fundamentals*.
- Appendix C on measuring the cutoff frequencies of a circuit.

MATERIALS

1	dual-trace oscilloscope
1	DMM
1	function generator
1	protoboard
5	resistors: 100 Ω, 1 kΩ, 10 kΩ, 47 kΩ, and 100 kΩ
1	2.2 nF capacitor
1	10 mH inductor

PROCEDURE

Part 1: Series LC Bandpass Filter

1. Measure and record the winding resistance of your coil.

$$R_W = \underline{\hspace{2cm}}$$

2. Perform the calculations listed below for the circuit shown in Figure 26.2. (Assume that $R_S = 50\ \Omega$, or refer to the manual for your function generator to obtain a more exact value.)

$$f_r = \frac{1}{2\pi\sqrt{LC}} = \underline{\hspace{3cm}}$$

$$X_L\ @\ f_r = \underline{\hspace{2.5cm}}$$

$$Q_L = \frac{X_L}{R_S + R_W + R_L} = \underline{\hspace{2.5cm}}$$

$$BW = \frac{f_r}{Q_L} = \underline{\hspace{2.5cm}}$$

3. Construct the circuit shown in Figure 26.2. Measure the resonant frequency of the circuit as follows:

   • Set the function generator to the value of f_r calculated in Step 2.

   • Vary the operating frequency around the calculated value of f_r until V_{RL} reaches its maximum value.

   Measure and record the frequency at which V_{RL} reaches its maximum value.

$$f_r = \underline{\hspace{3cm}}$$

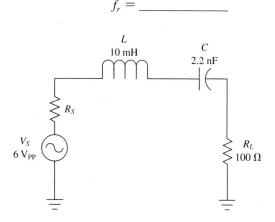

FIGURE 26.2   A series *LC* bandpass filter.

4. Measure the upper and lower cutoff frequencies of the circuit, and record your measurements below. Then, use these values to calculate the bandwidth of the circuit.

$$f_{C1} = \underline{\hspace{3cm}}$$

$$f_{C2} = \underline{\hspace{3cm}}$$

$$BW = \underline{\hspace{3cm}}$$

5. Use the values from Step 4 to calculate the loaded Q of this circuit.

$$Q_L = \frac{f_r}{\text{BW}} = \underline{\hspace{3cm}}$$

6. Change the load from 100 Ω to 1 kΩ. Repeat Steps 3 through 5, and record your measured and calculated values below.

$$f_r = \underline{\hspace{3cm}}$$

$$f_{C1} = \underline{\hspace{3cm}}$$

$$f_{C2} = \underline{\hspace{3cm}}$$

$$\text{BW} = \underline{\hspace{3cm}}$$

$$Q_L = \underline{\hspace{3cm}}$$

Part 2: Parallel LC Bandpass Filter

Note: The circuit shown in Figure 26.3 uses the same inductor and capacitor as in the previous circuit. Therefore, the two circuits have the same resonant frequency.

The series resistor (R_S) is added to prevent the function generator from loading down the circuit. If you were to connect the function generator directly to this circuit, the result would be a very low Q_L (as given in the Q_L equation).

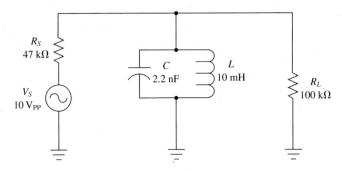

FIGURE 26.3 A shunt *LC* bandpass filter.

7. Refer to the circuit in Figure 26.3. Using the values of R_W, X_L, and f_r from Steps 1 and 2, calculate the following values:

$$Q = \frac{X_L}{R_W} = \underline{\hspace{3cm}}$$

$$R_P = Q^2 R_W = \underline{\hspace{3cm}}$$

$$Q_L = \frac{R_S \| R_P \| R_L}{X_L} = \underline{\hspace{3cm}}$$

$$\text{BW} = \frac{f_r}{Q_L} = \underline{\hspace{3cm}}$$

8. Construct the circuit shown in Figure 26.3. Measure the resonant and cutoff frequencies of the circuit, and record your measurements below.

$$f_r = \text{_____}$$

$$f_{C1} = \text{_____}$$

$$f_{C2} = \text{_____}$$

9. Using the values from Step 8, calculate the following values:

$$BW = \text{_____}$$

$$Q_L = \text{_____}$$

10. Change the load to 10 kΩ, and repeat Steps 8 and 9.

$$f_r = \text{_____}$$

$$f_{C1} = \text{_____}$$

$$f_{C2} = \text{_____}$$

$$BW = \text{_____}$$

$$Q_L = \text{_____}$$

QUESTIONS & PROBLEMS

1. Compare your calculated values in Step 2 with your measured values in Steps 3 through 5. Do these values agree? If not, explain any discrepancies. (*Hint:* Think about stray values.)

2. Refer to Step 6. What effect did increasing the load have on circuit BW and *Q*? Why did it have these effects?

3. Assume that a source with *high* output impedance was used to drive the circuit shown in Figure 26.2. What effect would this have on circuit BW and Q? Explain your reasoning.

4. Refer to Step 10. What effect did decreasing the load have on circuit BW and Q? Why did it have these effects?

5. Refer to the circuit shown in Figure 26.3. Assume that R_S was changed to 47 Ω. What effect would this have on circuit BW and Q? Explain your reasoning.

SIMULATION EXERCISE

Discussion

The Bode plotter allows you to quickly analyze the response curve of a bandpass filter. You can easily determine the resonant frequency, upper and lower cutoff frequencies, and thus the BW of the filter. When the Bode plotter is switched to the phase mode, you can also evaluate the phase response of the filter. At f_r the phase angle should be near 0° and ± 45° at the cutoff frequencies. This provides another way to determine the bandwidth of the filter.

1. Open file Ex26.1 from the Electronics Technology Fundamentals companion web site (www.prenhall.com/paynter). This file contains two circuits. They are the same circuits as those shown in Figures 26.2 and 26.3. We have assumed that $R_W = 20\ \Omega$.

2. Note that the settings on the two Bode plotters are slightly different. Both circuits are swept from 20 kHz to 75 kHz, but the vertical scale for the series filter is set to 0 to -10 dB, while the vertical scale for the shunt filter is 0 to -30 dB.

3. Open both Bode plotters and run the simulation until you get a trace on both plotters.

4. Use the cursor to determine the following values for the series filter: (Remember that the cutoff frequencies are 3 dB lower than the center frequency.)

   $f_r =$ _____ $f_{C1} =$ _____ $f_{C2} =$ _____

5. Use your results from Step 4 to calculate the following values:

   $BW =$ _____ $Q_L =$ _____

6. Switch the Bode plotter mode to "phase". Use the cursor, and your results from Step 4, to determine the phase of the filter at the resonant frequency and both cutoff frequencies. Record these values below.

   $\theta @ f_r =$ _____ $\theta @ f_{C1} =$ _____ $\theta @ f_{C2} =$ _____

7. Repeat Steps 3 through 6 for the parallel shunt filter.

   $f_r =$ _____ $f_{C1} =$ _____ $f_{C2} =$ _____

   $BW =$ _____ $Q_L =$ _____

   $\theta @ f_r =$ _____ $\theta @ f_{C1} =$ _____ $\theta @ f_{C2} =$ _____

Questions

1. Compare your results in the simulation portion of the exercise with those in the hardware section. Were there any significant variations between the two parts of the exercise? What factors do you think would contribute to these variations?

2. Refer to the phase characteristics of the two circuits. In both cases, the phase angle is positive below resonance and negative above resonance. Explain why this is the case, given that one circuit is a series resonant circuit and the other is a parallel resonant circuit.

Exercise 27

RL and RC Switching Circuit Pulse Response

OBJECTIVES

After completing this exercise, you should be able to:

- Analyze the pulse response of both *RL* and *RC* switching circuits.
- Draw the waveform for either component in an *RL* or *RC* switching circuit.
- Analyze the frequency response of both *RL* and *RC* switching circuits.

DISCUSSION

As you probably know, *digital* circuits employ square and rectangular (collectively known as *pulse*) waveforms. The pulse response of *RC* and *RL* circuits is quite a bit different than their response to a sinusoidal input. In this exercise, you will look at the response of *RL* and *RC* circuits to a square wave input.

In the first part of this exercise, you will focus on the pulse response of *RL* circuits. In the second part of this exercise, you will use an *RC* circuit to focus on the frequency response of switching circuits.

LAB PREPARATION

Review Chapter 16 of *Electronics Technology Fundamentals*.

MATERIALS

1   dual-trace oscilloscope
1   function generator
1   protoboard
1   100 Ω resistor
1   0.1 μF capacitor
1   1 mH inductor

PROCEDURE

Part 1: RL Circuit Pulse Response

1.   Calculate the time constant (τ) for the circuit shown in Figure 27.1, and record the value below. Then, solve for the amount of time required for the circuit current to complete its rise (or decay), and record this value below.

$$\tau = \underline{\hspace{4cm}}$$

$$5\tau = \underline{\hspace{4cm}}$$

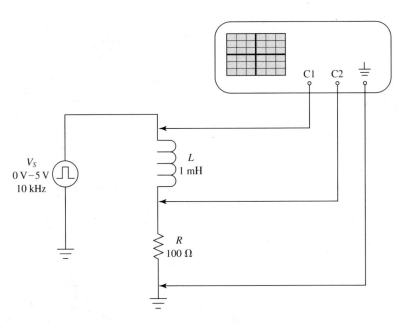

FIGURE 27.1   An *RL* pulse circuit.

2.   Construct the circuit shown in Figure 27.1. Channel 1 (C1) is connected to display V_S. Channel 2 (C2) is connected to display V_R. The oscilloscope should be set up as follows:
   • Both oscilloscope inputs should be dc coupled.
   • Set the vertical sensitivity to 2 V/div (both channels).
   • Set the time base to 20 μs/div.
   • Adjust the vertical position of the two waveforms so that they are displayed with the input waveform above V_R.

3.   Draw the V_S and V_R waveforms in Figure 27.2.

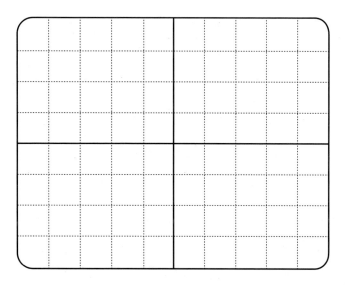

Vertical sensitivity: _____ Time base: _____

FIGURE 27.2

4. Change the oscilloscope settings as follows:
   • Vertical sensitivity = 1 V/div
   • Time base = 10 μs/div
   Determine the time needed for the rise curve of the waveform to reach 5 V (5τ), and record this value below. Use this result to calculate the time constant of this circuit.

$$5\tau = \underline{\hspace{2cm}}$$

$$\tau = \underline{\hspace{2cm}}$$

5. Carefully adjust the time-base calibration control so that one complete rise curve occupies exactly five major divisions on the horizontal scale. (This effectively divides the waveform into five equal portions, or five time constants.) Measure, as accurately as possible, the voltage of the rise curve at each time constant, and record these values in Table 27.1.

6. Reverse the order of the components so that Channel 2 is measuring V_L, and repeat Step 2 for the V_L waveform. Draw the source and V_L waveforms in Figure 27.3.

7. Repeat Step 5 for the V_L waveform, but this time, measure the voltages on the decay curve. Record your measurements in Table 27.2.

TABLE 27.1   Rise Curve Measurements

Time	V_R
τ_1	
τ_2	
τ_3	
τ_4	
τ_5	

TABLE 27.2   Decay Curve Measurements

Time	V_L
τ_1	
τ_2	
τ_3	
τ_4	
τ_5	

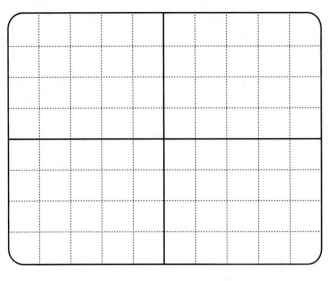

Vertical sensitivity: _____ Time base: _____

FIGURE 27.3

Part 2: RC Circuit Pulse Response

8. Calculate the time constant for the circuit shown in Figure 27.4, and record this value below.

$$\tau = \text{_____}$$

9. Construct the circuit shown in Figure 27.4. Set the oscilloscope time base so that you see approximately one complete cycle on the display. Determine the time constant of the circuit, and record your results below. Then, draw the V_S and V_C waveforms in Figure 27.5a.

$$5\tau = \text{_____}$$

$$\tau = \text{_____}$$

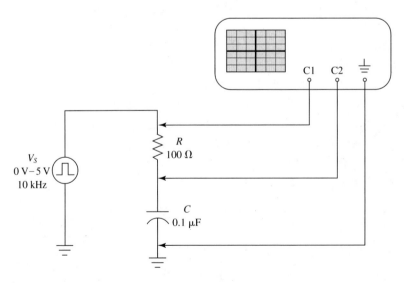

FIGURE 27.4 An *RC* pulse circuit.

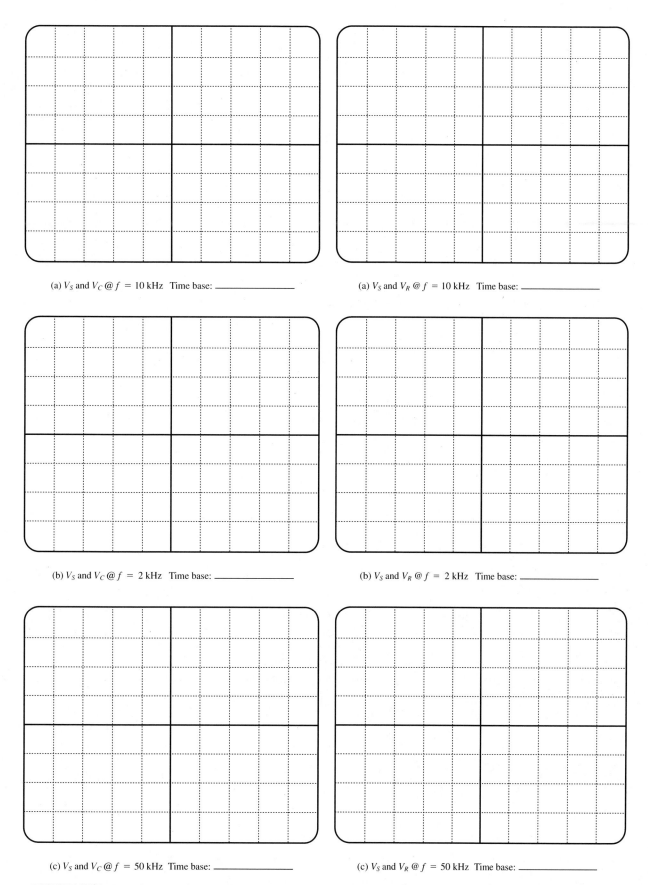

(a) V_S and V_C @ $f = 10$ kHz   Time base: _____

(a) V_S and V_R @ $f = 10$ kHz   Time base: _____

(b) V_S and V_C @ $f = 2$ kHz   Time base: _____

(b) V_S and V_R @ $f = 2$ kHz   Time base: _____

(c) V_S and V_C @ $f = 50$ kHz   Time base: _____

(c) V_S and V_R @ $f = 50$ kHz   Time base: _____

FIGURE 27.5 V_S and V_C waveforms at three frequencies.

FIGURE 27.6 V_S and V_R waveforms at three frequencies.

10. Change the input frequency to 2 kHz, and adjust the oscilloscope time base so that you again see approximately one complete cycle on the display. Then, draw the V_S and V_C waveforms in Figure 27.5b.

11. Change the input frequency to 50 kHz, and adjust the oscilloscope time base so that you again see approximately one complete cycle on the display. Then, draw the V_S and V_C waveforms in Figure 27.5c.

12. Reverse the order of the components, and measure V_R for the same three input frequencies. Then, draw the waveforms in Figure 27.6.

QUESTIONS & PROBLEMS

1. Refer to Steps 1 and 4. Do your measured and calculated values for τ agree? If not, what do you think caused the discrepancy?

2. Refer to Step 2 and the V_R waveform in Figure 27.2. Explain the shape of the waveform. Base your explanation on the characteristics of inductors.

3. Refer to Steps 5 and 7. Calculate the percent of change from one time constant to the next for those results. Does the percent of change agree with the theoretical value of 63.2%? If not, what do you think caused the discrepancy?

4. Refer to the V_R and V_C waveforms in Figures 27.5 and 27.6. Explain what happens when the operating frequency is much lower (or higher) than the time that it takes for the capacitor to fully charge or discharge.

5. Based on your answer to Question 4, explain how you think the V_L and V_R waveforms would look for the circuit shown in Figure 27.1 at the same frequencies (2 kHz and 50 kHz). Feel free to sketch both waveforms.

SIMULATION EXERCISE

Discussion

The instruments in the simulator provide opportunities to ask—and to answer—measurement questions that are either difficult or time-consuming when done using hardware. Many oscilloscopes provide for *cursor-based* measurements, and this feature has been incorporated into the simulated scope. This exercise will show you how to use some of these features.

One of the problems with using an oscilloscope is that it can only measure voltages with respect to ground. That is, if neither end of the component of interest is grounded, then the circuit must be rewired so that one end of that component is connected to ground. This problem can be eliminated in hardware by operating the oscilloscope in the *differential* mode. The simulator solution is to use a VCVS (as we have done in previous labs). As you may recall, the VCVS is essentially a voltage source that generates an output (referenced to ground) that is some multiple of a floating input voltage.

Procedure

1. Open file Ex27.1 from the Electronics Technology Fundamentals companion web site (www.prenhall.com/paynter). This file contains two circuits. They are similar to those shown in Figures 27.1 and 27.4. The only difference is the addition of the VCVS to each circuit, and the component order in the *RC* circuit. We will begin with the *RL* pulse circuit.

2. Run the simulation, and use the single (Sing.) function in the time-base control section of the oscilloscope to capture one cycle of the waveform. Note that after the single display has been captured, time base and vertical sensitivity values can still be changed.

3. Position cursor 1 on the rising edge of V_S and cursor 2 on the falling edge. The [T2 − T1] display should read very close to 50 μsec. If not, recheck your settings.

4. Move cursor 2 closer to cursor 1 until the [T2 − T1] reading is as close to 40 μsec as possible. Read and record the value of V_R (shown in the VB2 display) in Table 27.3.

TABLE 27.3   **Simulation Results**

T2–T1	V_R	V_L
40 μs		
30 μs		
20 μs		
10 μs		

5. Repeat the cursor 2 movement, and record the VB2 readings at [T2 − T1] values of 30 μs, 20 μs, and 10 μs in Table 27.3.

6. Disconnect the Channel 2 lead from V_{R1}, and reconnect it to the VCVS output.

7. Set the vertical sensitivity of Channel 1 to 5 V/div. (Leave Channel 2 at 2 V/div.)

8. Adjust the vertical position of the two waveforms so that they are displayed with the input waveform above V_{R1} as follows:

    • Set the Channel 1 Y position to 2.

    • Set the Channel 2 Y position to −1.

9. Repeat Steps 3 through 7, measuring and recording the value for V_L (as shown in VB2).

10. Refer to the *RC* pulse circuit. Compare the V_S waveform with the V_R waveform at the following frequencies:

    • 10 kHz.

    • 2 kHz.

    • 50 kHz.

    Draw the waveforms in Figure 27.7.

11. Repeat Step 10, this time comparing the V_S waveform with the V_C waveform at the frequencies listed. Draw the waveforms in Figure 27.8.

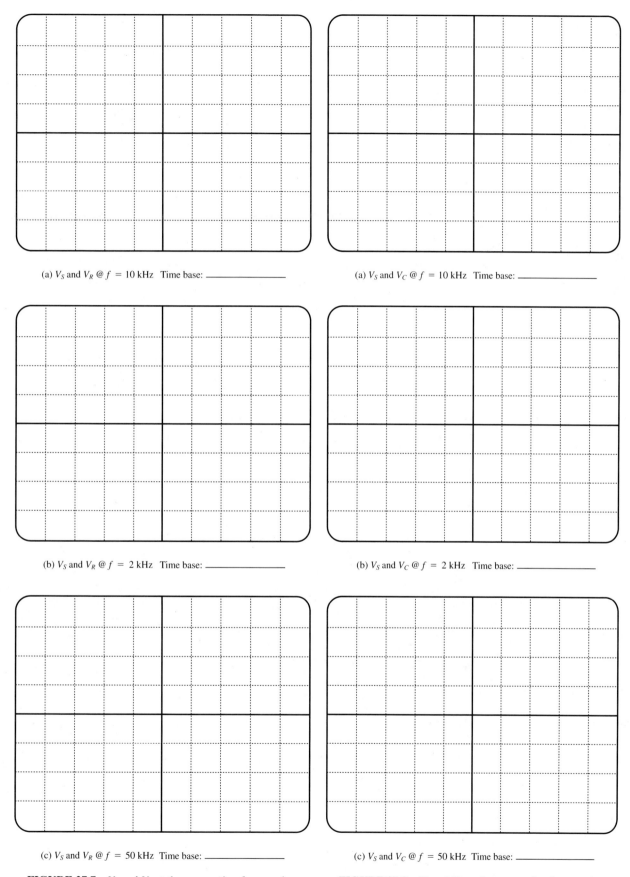

(a) V_S and V_R @ $f = 10$ kHz Time base: _____

(a) V_S and V_C @ $f = 10$ kHz Time base: _____

(b) V_S and V_R @ $f = 2$ kHz Time base: _____

(b) V_S and V_C @ $f = 2$ kHz Time base: _____

(c) V_S and V_R @ $f = 50$ kHz Time base: _____

(c) V_S and V_C @ $f = 50$ kHz Time base: _____

FIGURE 27.7 V_S and V_R at three operating frequencies.

FIGURE 27.8 V_S and V_C at three operating frequencies.

Question

1. Explain how the waveforms in this exercise demonstrate that each simulation circuit complies with Kirchhoff's voltage law.

Part 3
Electronic Devices

Exercise 28

Diode Characteristics

OBJECTIVES

After completing this exercise, you should be able to:

- Measure the forward voltage across a diode, and determine if the component is faulty.
- Demonstrate the forward current and voltage characteristics of *pn*-junction and zener diodes.
- Demonstrate the reverse current and voltage characteristics of *pn*-junction and zener diodes.

DISCUSSION

A diode can be tested using a simple ohmmeter, but this is not a very accurate test. Most modern multimeters have a diode-check function, which allows you to determine the actual voltage across a forward-biased diode. If the forward voltage falls within its expected range, the diode is considered good. If the diode test indicates the component is shorted or open, then the diode should be replaced.

A diode is *forward biased* when its anode is more positive than its cathode. As *forward current* (I_F) increases, so does *forward voltage* (V_F) across the device. However, V_F increases at a much lower rate than I_F, because the forward resistance of a diode *decreases* as I_F increases. Therefore, V_F increases at a very low rate when a diode is operated above its *knee voltage* (V_K). This is true of *pn*-junction diodes, zener diodes, and even LEDs.

A diode is *reverse biased* when its cathode is more positive than its anode. When a *pn*-junction diode is reverse biased, the *reverse current* (I_R) through the device is extremely low, even with a significant reverse voltage applied. This is not necessarily the case with a *zener diode.* Zener reverse current remains low until V_R reaches the *zener voltage* (V_Z)

rating of the component. When the magnitude of V_R reaches V_Z, I_R increases abruptly. The forward and reverse characteristics of *pn*-junction and zener diodes are the focus of the second part of this exercise.

LAB PREPARATION

Review Chapter 17 of *Electronics Technology Fundamentals*.

MATERIALS

1   variable dc power supply
2   DMMs (one with diode-check function)
1   protoboard
2   resistors: 100 Ω and 1 kΩ
1   1 kΩ potentiometer
1   1N4148 small signal diode
1   1N5240 zener diode

PROCEDURE

Part 1: Diode Testing

Note: The schematic and component symbols for *pn*-junction and zener diodes are shown in Figure 28.1. Note that the indicator band on the component is always closest to the cathode terminal.

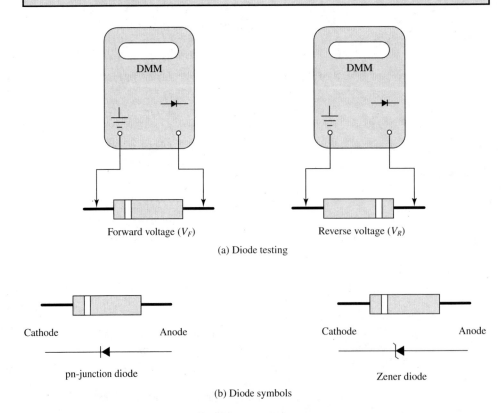

FIGURE 28.1   Diode testing.

1. Set your DMM to the diode test position.
2. Connect the DMM to the 1N4148 diode as shown in Figure 28.1a. Measure the forward voltage (V_F) across the diode, and record your measurement in Table 28.1.

TABLE 28.1   Diode Test Measurements

Diode	V_F	V_R
1N4148		
1N5240		

3. Reverse the diode as shown in Figure 28.1a. Measure its reverse voltage (V_R), and record this reading in Table 28.1.
4. Repeat Steps 2 and 3 for the 1N5240 zener diode. Record the meter readings in Table 28.1.

Part 2: Diode Voltage and Current Characteristics

5. Construct the circuit shown in Figure 28.2. The voltage source should be set initially to 10 V. The potentiometer (R_1) is used to adjust the current through the diode. R_2 limits the diode current to a value lower than the rated maximum forward current.

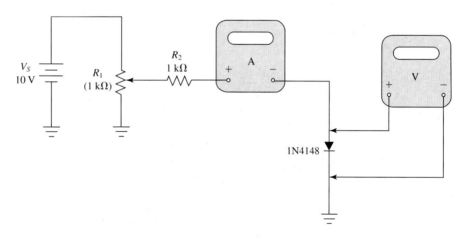

FIGURE 28.2

6. Adjust R_1 so that $I_F = 0.5$ mA (500 µA). Measure V_F, and record this value in Table 28.2.
7. Repeat Step 6 for all the current values listed in Table 28.2.
8. Replace the 1N4148 with the 1N5240 zener diode and repeat steps 6 & 7.

TABLE 28.2   Diode Forward Currents and Voltages

I_F (mA)	V_F (1N4148)	V_F (1N5240)
0.5		
1.0		
1.5		
2.0		
2.5		
3.0		
3.5		
4.0		
4.5		
5.0		

9. Use your results from Table 28.2 to plot the I_F versus V_F curve for both diodes in Figure 28.3.

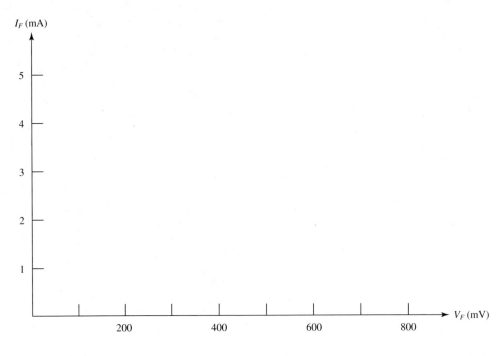

FIGURE 28.3 I_F versus V_F.

10. Construct the circuit shown in Figure 28.4. Note that the 1N4148 diode is now reverse biased. Adjust R_1 until the voltage across the diode measures 12 V.

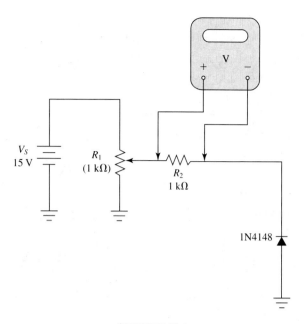

FIGURE 28.4

11. Measure V_{R2}, and record this value below. Use this value (and the rated value of R_2) to calculate I_{R2}. This equals the reverse current through the diode.

$$V_{R2} = \underline{\hspace{3cm}} \qquad I_{R2} = \underline{\hspace{3cm}}$$

12. Construct the circuit shown in Figure 28.5. Adjust V_S until $I_Z = 1.0$ mA.

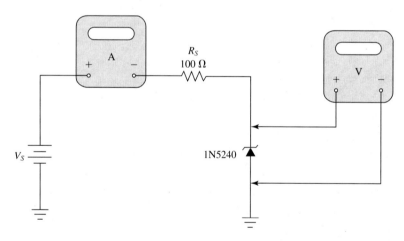

FIGURE 28.5 Zener diode test circuit.

13. Measure the voltage across the zener diode, and record this value in Table 28.3. Repeat this procedure for all the current values listed in the table.

TABLE 28.3   Zener Voltage Measurements

I_Z (mA)	V_Z	I_Z (mA)	V_Z
1.0		10	
2.0		15	
3.0		20	
4.0		25	
5.0		30	

14.   Use the values from Table 28.3 to plot the I_Z versus V_Z curve in Figure 28.6.

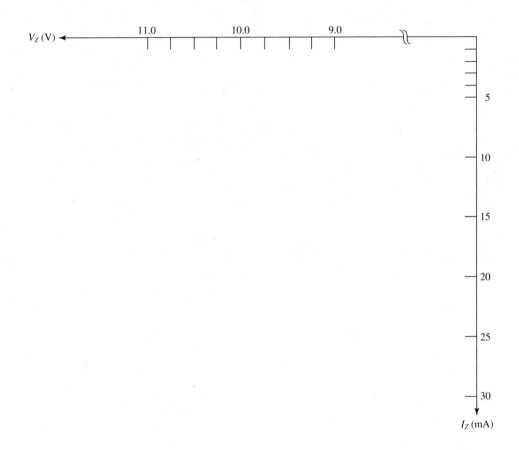

FIGURE 28.6   Zener reverse operating curve.

1. Based on your knowledge of *pn* junctions, what do you think would happen to the measurements that you made in Steps 2 and 3 if the diodes were heated to 50°C? Explain your reasoning.

2. Refer to your graphs in Figure 28.3. How did the forward characteristics of the two diodes compare? Explain why they are so similar (or dissimilar).

3. Refer to your results in Step 11. Was the reverse current in the range that you expected? What would happen to the reverse current if the diode was heated to 50°C? Explain your reasoning.

4. Refer to your results in Table 28.3. Explain how you could use these results to calculate the zener impedance of this 1N5240 zener diode. Then, use this technique to calculate Z_Z.

SIMULATION EXERCISE

Discussion

Diodes are used in a wide variety of applications. It should not be too surprising that different application diodes should have different characteristics. When analyzing a diode circuit, we tend to assume that $V_F \cong 700$ mV. This is not always the case.

A diode can be tested in an active state (with power applied) or in a passive state. When testing with power applied we use a voltmeter. When we test a diode with no power applied, we use a diode checker. As you will see, these two methods do not yield the same results.

In this exercise you will measure the value of V_F for 4 different diodes. The 1N4001 is a general purpose rectifier diode. (You will study rectifier circuits in Exercise 29.) The 1N4148 is a high speed switching diode. This diode has a rated recovery time of just 4 ns. The 1N4944 is another rectifier diode, but with faster switching capabilities than the 1N4001. Finally, the 1N5395 is an industrial rectifier diode capable of handling much higher currents than the 1N4001 rectifier diode. In this simulation exercise you will look more closely at the forward characteristics of these diodes.

Procedure

1. Open file Ex28.1 from the Electronics Technology Fundamentals companion web site (www.prenhall.com/paynter). In this exercise you will be using the Agilent multimeter as both a voltmeter and a diode checker. Note the switch labeled as SW1. When the switch is open, there is no power applied to the diode. This allows us to make a *passive* measurement of diode forward voltage using the DMM as a diode checker. With SW1 closed, power is applied to the diode and we must use the DMM as a voltmeter to measure V_F.

2. Run the simulation and double-click on the Agilent DMM. Click on the power button and note that the display tells you that the meter is set to measure dc volts. To set the meter to the diode-test mode, click on the "shift" button and then the "Cont" button with the diode symbol above it. Measure the value of V_F for the 1N4001 using the diode-test function and record this value in Table 28.4.

3. Change the setting of the DMM to measure dc volts and close SW1. Measure the value of V_F and record this value in Table 28.4.

TABLE 28.4 V_F Measurements

Diode	V_F Diode Checker	V_F Voltmeter
1N4001		
1N4148		
1N4944		
1N5395		

4. Repeat Steps 2 and 3 for the other diodes listed in Table 28.4. The easiest way to change diodes is to double-click on the diode and choose "Replace". Make certain that SW1 is open for all tests using the diode-test function and closed for tests using the voltmeter.

5. Now, using the 1N5395, double-click on the diode and fault the diode open. Measure the value of V_F using both measurement techniques and record these values below.

$$V_F \text{ (open)} = \underline{\hspace{2cm}} \text{ Diode Check}$$

$$V_F \text{ (open)} = \underline{\hspace{2cm}} \text{ Voltmeter}$$

6. Fault the diode as a short and repeat Step 5.

$$V_F \text{ (short)} = \underline{\hspace{2cm}} \text{ Diode Check}$$

$$V_F \text{ (short)} = \underline{\hspace{2cm}} \text{ Voltmeter}$$

Questions

1. The value of V_F for each diode that you measured using the diode checker should be well below the value measured using the voltmeter. Given the information that this diode checker has a rated current of 1 mA, explain why this discrepancy exists.

2. Explain your results for the open and shorted diode tests using the two meters. Why do you get very different results for the open diode, but the same results for the shorted diode?

Exercise 29

Diode Rectifier Circuits

OBJECTIVES

After completing this exercise, you should be able to:

- Demonstrate the strengths and weaknesses of the three basic rectifier circuits.
- Draw the output waveforms for the three basic rectifier circuits.
- Demonstrate the effect of filtering on rectifier circuits.

DISCUSSION

The three basic rectifier configurations are the half-wave, full-wave, and bridge rectifiers. The output of a positive half-wave rectifier is shown in Figure 29.1a. Figure 29.1b shows the output of a positive full-wave or bridge rectifier. Since the bridge rectifier is by far the most common, you will examine the effect of filters on this rectifier only.

This is the first exercise in which you will be working with relatively high voltages. The transformer in this exercise has a 120 V_{ac} input (line voltage). Be very careful whenever you are working with any electrical circuit—but especially so when working with higher voltages and/or currents.

LAB PREPARATION

- Review Sections 18.1 through 18.4 of *Electronics Technology Fundamentals*.
- Review the discussion on fault analysis in Appendix D.
- Make eight copies of the *fault analysis chart* in Appendix D (for the fault simulations portion of the exercise).

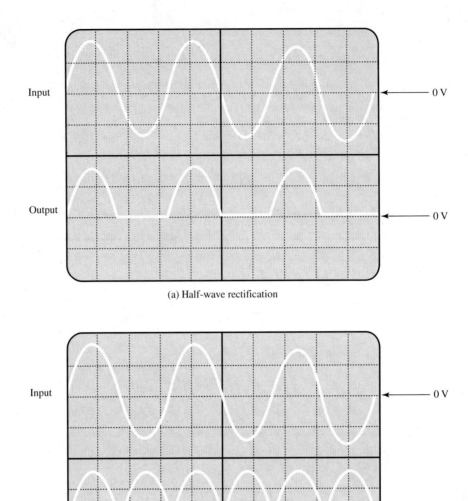

(a) Half-wave rectification

(b) Full-wave rectification

FIGURE 29.1 Rectifier input and output waveforms.

MATERIALS

1	dual-trace oscilloscope
1	DMM
1	power transformer: 25.2 V secondary, center-tapped
1	protoboard
1	5.6 kΩ resistor
2	electrolytic capacitors: 10 μF and 100 μF (both rated at 50 V)
4	1N4001 rectifier diodes
1	¼-amp fuse and fuse holder

PROCEDURE

Part 1: Half- and Full-Wave Rectifiers

1. Construct the half-wave rectifier circuit shown in Figure 29.2. The input to Channel 2 (C2) should be set for dc coupling. Use Figure 29.1a as a model for positioning the input and output waveforms on the screen.

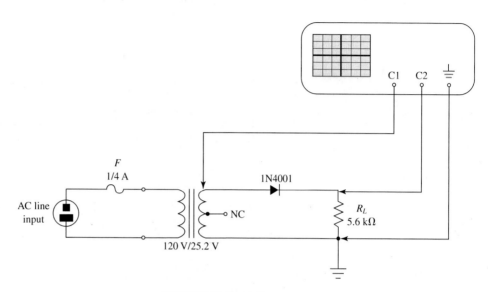

FIGURE 29.2   Half-wave rectifier.

2. Draw the circuit input and output waveforms in Figure 29.3a.

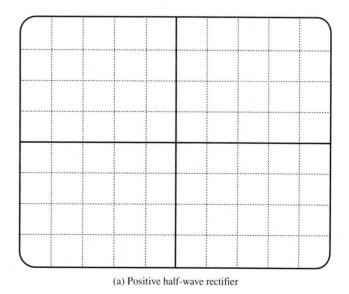

(a) Positive half-wave rectifier

FIGURE 29.3   Half-wave rectifier waveforms. *(continues)*

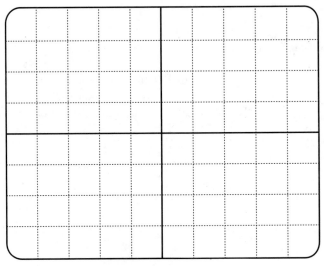

(b) Negative half-wave rectifier

FIGURE 29.3 *(continued)*

3. Use the oscilloscope to measure $V_{S(pk)}$ and use the DMM to measure the dc voltage across the load (V_{ave}). Record these values below.

$$V_{S(pk)} = \underline{\hspace{3cm}} \qquad V_{ave} = \underline{\hspace{3cm}}$$

4. Reverse the diode. Measure and draw both waveforms in Figure 29.3b. Measure V_{ave} with your DMM, and record this value below.

$$V_{ave} = \underline{\hspace{3cm}}$$

5. Construct the full-wave rectifier circuit shown in Figure 29.4.

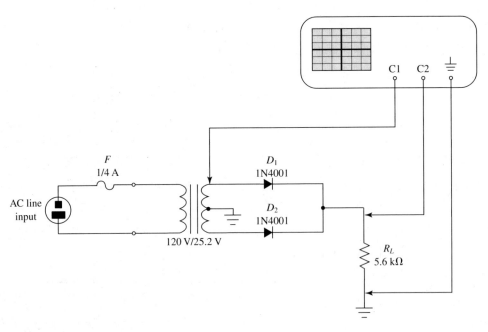

FIGURE 29.4 Full-wave rectifier.

6. Draw the circuit waveforms in Figure 29.5a. Then, use your DMM to measure the dc voltage (V_{ave}) across R_L. Record this value below.

$$V_{ave} = \underline{\hspace{2cm}}$$

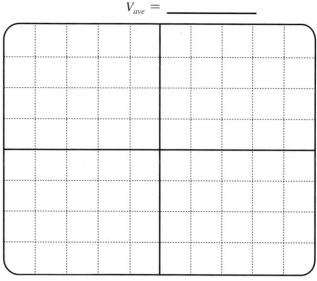

(a) Positive full-wave rectifier

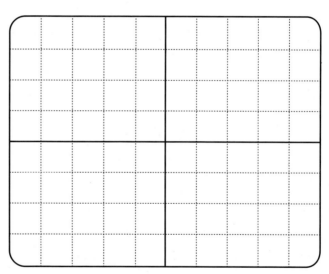

(b) Negative full-wave rectifier

FIGURE 29.5   Full-wave rectifier waveforms.

7. Reverse the direction of the diodes, and draw the circuit waveforms in Figure 29.5b. Then, use your DMM to measure the dc load voltage (V_{ave}). Record this value below.

$$V_{ave} = \underline{\hspace{2cm}}$$

8. Construct the circuit shown in Figure 29.6, and draw the circuit waveforms in Figure 29.7. Then, measure the dc load voltage (V_{ave}), and record this value below.

$$V_{ave} = \underline{\hspace{2cm}}$$

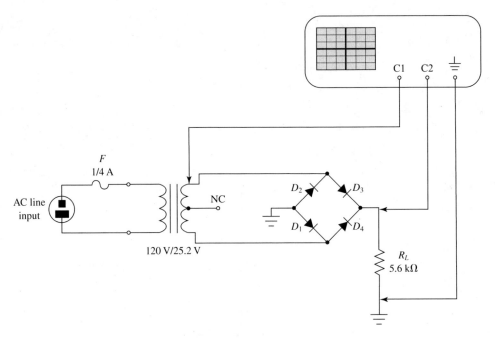

FIGURE 29.6 A bridge rectifier.

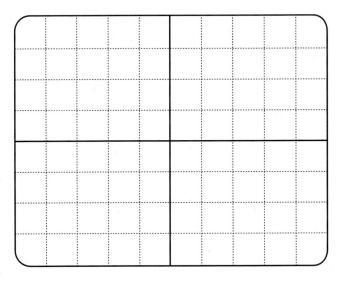

FIGURE 29.7 Bridge rectifier waveforms.

Part 2: Rectifier Filtering

As you have seen in the first part of this exercise, the output from a rectifier is a pulsating dc voltage. The filter in a linear power supply is designed to reduce the variations in this dc voltage. As you will see shortly, the value of the filter capacitor is one of the factors that determine how effective the filter is. No filter is perfect, however, so the variations in the dc voltage are never completely eliminated. The remaining variations in the dc voltage are referred to as the *ripple voltage* (V_r).

9.   Add a 10 μF capacitor in parallel with the bridge rectifier load as shown in Figure 29.8. Use the DMM to measure V_{ave}. Record this value below.

$$V_{ave} = \underline{\hspace{2cm}}$$

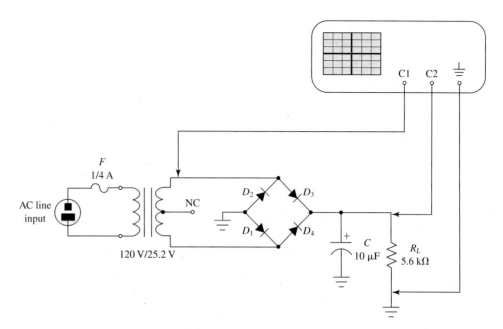

FIGURE 29.8   A filtered bridge rectifier.

Note: It is extremely important to observe proper polarity when working with electrolytic capacitors. If installed backwards, they will fail and may explode.

10. Use the oscilloscope to observe and measure the ripple voltage. (*Note:* The Channel 2 input must be ac coupled to measure the ripple voltage.) Draw the ripple waveform on the upper portion of the graph in Figure 29.9, and record its measured peak-to-peak value.

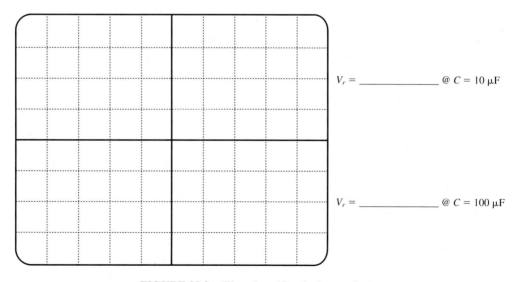

$V_r =$ _____ @ $C = 10\ \mu F$

$V_r =$ _____ @ $C = 100\ \mu F$

FIGURE 29.9   Filtered rectifier ripple waveforms.

11. Change the filter capacitor from 10 μF to 100 μF. Power up, and repeat Steps 9 and 10. Draw the ripple voltage waveform on the lower portion of the graph in Figure 29.9, and record its measured peak-to-peak value.

$$V_{ave} = \text{_____}$$

1. Explain why the peak secondary voltage in the full-wave rectifier was so much lower than in the other two circuits.

2. Use the appropriate relationships to calculate V_{ave} for each circuit. Compare these values with your results in Steps 3, 6, and 8. Explain any differences between the measured and the calculated values.

$$V_{ave} \text{ (half-wave)} = \underline{\hspace{2cm}}$$

$$V_{ave} \text{ (full-wave)} = \underline{\hspace{2cm}}$$

$$V_{ave} \text{ (bridge)} = \underline{\hspace{2cm}}$$

3. Refer to Figure 29.6. Explain how you would modify this circuit to obtain a negative load voltage.

4. Refer to your answer to Question 3. What precaution would be required when adding a filter capacitor to this circuit?

Discussion

In this simulation you will begin to hone your troubleshooting skills. In most real-world situations, you will have to troubleshoot a circuit using only an oscilloscope and a multimeter. Knowing which measurement to make, and being able to interpret what each reading tells you about the fault, is a skill that you will develop with experience.

There are two approaches to teaching the art of troubleshooting. One way is to present the student with a circuit with an unknown fault, and then you try to determine what the fault is. The other method is to introduce a specific fault into a known circuit. This allows the student to become familiar with how a particular fault presents itself. This is the approach that we will use in this exercise. You will be asked to introduce specific faults into several rectifier circuits, determine the primary symptoms of the fault, and then interpret these symptoms.

Example: A full-waver rectifier may have low output voltage as measured with a voltmeter. A scope check might reveal high ripple voltage. The oscilloscope measurement of the ripple frequency would show that the circuit is operating as a half-wave rectifier. An analysis of these faults would lead you to determine that there is an open diode or transformer secondary.

Fault Simulation

Note: Some faults that are inserted using the simulator can be dangerous to equipment and/or personnel when they occur in real circuits. Simulations provide exposure to such faulyts in a safe environment.

Procedure

1. Open files Ex29.1 and Ex29.2 from the Electronics Technology Fundamentals companion web site (www.prenhall.com/paynter). When you open the files, note the tabs on the lower left-hand portion of your screen. By clicking on these tabs, you can view either of the two files that you have opened.
2. Run simulations on all three circuits and make sure that each circuit is working properly before you begin to introduce any faults.
3. Open the transformer primary (T1) of the half-wave rectifier circuit, and run the simulation. Using the oscilloscope and multimeter supplied, determine the major symptoms of the failure. Record these symptoms (and any others that you feel are pertinent) in one of the fault analysis charts that you have copied from Appendix D. Then, explain why the circuit responded to the fault in the way that it did, in the analysis section of the chart. Identify this analysis as Fault 29-1.
4. Run each of the fault simulations that follow. Record your results/analysis for each fault in one of the fault analysis charts.

Fault Simulations

Fault 29-2   Shorted transformer secondary (T_3 to T_4) in the half-wave rectifier.

Fault 29-3   Shorted diode in the half-wave rectifier.

Fault 29-4   Open diode in the full-wave rectifier.

Fault 29-5   Open transformer secondary (T_4) in the full-wave rectifier.

Fault 29-6   Shorted transformer secondary center tap (T_5) in the full-wave rectifier.

Fault 29-7   Open D_3 in the bridge rectifier.

Fault 29-8   Open D_4 in the bridge rectifier.

Exercise 30

Clippers, Clampers, and Voltage Multipliers

OBJECTIVES

After completing this exercise, you should be able to:

- Demonstrate the operation of the shunt clipper circuit.
- Demonstrate the operation of the diode clamper circuit.
- Measure the dc offset of a signal with an oscilloscope.
- Demonstrate the operation of a half-wave voltage multiplier.

DISCUSSION

The half-wave rectifier is a simple *series* clipper. It "clips" either the positive or the negative alternation of its input waveform, depending on the polarity of the diode. Since you examined series clipper operation in the rectifier lab, you will focus on the *shunt* clipper in this exercise.

The *clamper* is a diode circuit used to change the *dc reference* of a waveform *without significantly altering the shape of that waveform*. The *positive* clamper shifts its input waveform in the positive direction; the *negative* clamper shifts it in the negative direction. The negative clamper is identical to the positive clamper except for the polarity of the diode and capacitor. Therefore, you will only investigate the positive clamper in this exercise.

A *voltage multiplier* is a diode circuit used to provide a dc output that is a specified multiple of the peak value of its input signal voltage. For example, the dc output from a voltage *doubler* is approximately *two times* its peak input voltage. The voltage *tripler* provides a dc output that is approximately *three times* its peak input voltage, and so on. There is a slight voltage across each diode in the multiplier circuit, so the output voltage cannot truly reach the design multiple of the peak input voltage. In this exercise, you will investigate the operation of a half-wave voltage doubler.

- Review Sections 18.5 through 18.7 of *Electronics Technology Fundamentals*.
- Review the discussion on fault analysis in Appendix D.
- Make five copies of the *fault analysis chart* in Appendix D (for the fault simulations portion of the exercise).

MATERIALS

1	dual-trace oscilloscope
1	function generator
1	protoboard
2	resistors: 1 kΩ and 100 kΩ
3	capacitors: 1 μF and 47 μF (2) (25 V rated)
2	1N4001 diodes

PROCEDURE

Part 1: The Shunt Clipper

1. Construct the circuit shown in Figure 30.1. Channel 1 (C1) is connected to display V_S, and Channel 2 (C2) is connected to display V_L.

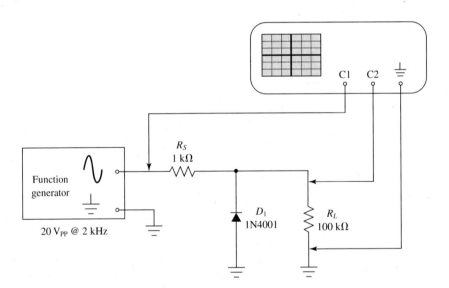

FIGURE 30.1 A shunt clipper.

2. Draw the input and output waveforms in Figure 30.2a.
3. Reverse the direction of D_1 and repeat Step 2. Draw the two waveforms in Figure 30.2b.

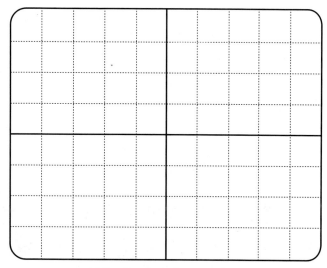

(a) Negative shunt clipper waveforms

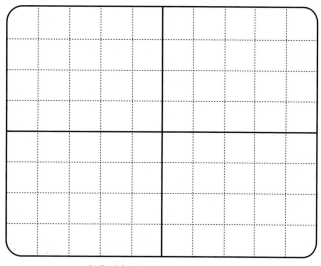

(b) Positive shunt clipper waveforms

FIGURE 30.2   Shunt clipper input and output waveforms.

Part 2: Clampers

4. Construct the circuit shown in Figure 30.3. Note the direction of the diode and the polarity of the capacitor. Channel 1 (C1) is connected to display V_S, and Channel 2 (C2) is connected to display V_L. Both channel inputs should be dc coupled.
5. Draw the input and output waveforms in Figure 30.4a.
6. Reverse the diode and the capacitor and repeat Step 4. Draw the two waveforms in Figure 30.4b.

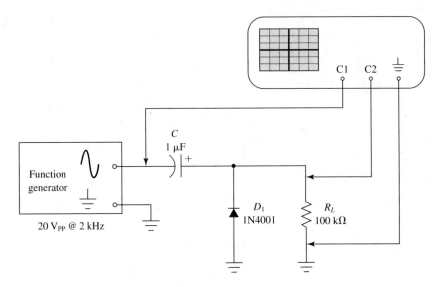

FIGURE 30.3 A clamper.

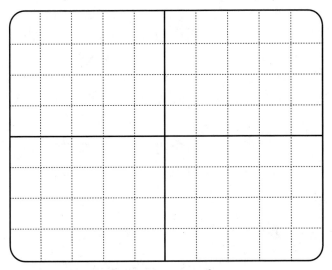

(a) Positive clamper waveforms

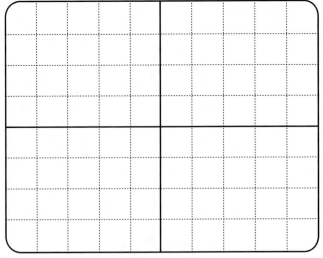

(b) Negative clamper waveforms

FIGURE 30.4 Clamper waveforms.

7. There is an easy way to measure the *dc average* (dc offset) of any waveform using the oscilloscope:

- Set the Channel 2 input to ac coupling. Note the position of the positive (or negative) peak of the waveform.

- While viewing the waveform, switch the channel input to dc coupling. Note the direction of the shift in the position of the waveform and the number of divisions by which it shifts. If it shifts *up* on the display, it is a *positive* offset. If it shifts *down,* it is a *negative* offset.

- Multiply the number of divisions by the vertical sensitivity (V/div) setting. The result is the dc average (dc offset) of the waveform.

Use this technique on the circuit from Step 6, and record the dc average of the output waveform below.

$$V_{ave} = \underline{\hspace{3cm}}$$

Part 3: Voltage Multipliers

8. Construct the circuit shown in Figure 30.5. *It is very important that the polarity of the two capacitors be correct.*

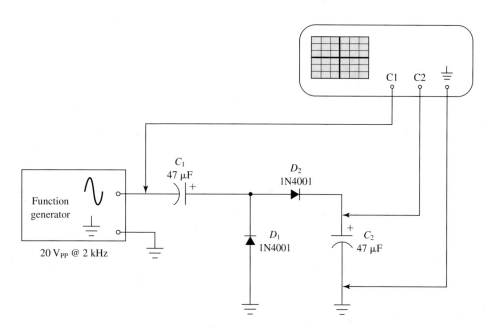

FIGURE 30.5   A half-wave voltage doubler.

9. Use your DMM to measure the dc output voltage (V_{dc}), and record this value below. Then, set the oscilloscope to ac coupling, and measure the output ripple voltage. Record this value below as well.

$$V_{dc} = \underline{\hspace{3cm}} \qquad V_r = \underline{\hspace{3cm}}$$

QUESTIONS & PROBLEMS

1. What purpose does R_S serve in the circuit shown in Figure 30.1?

2. Refer to Figure 30.2. Was the input waveform clipped at *exactly* 0 V? If not, explain why.

3. Refer to Step 7. There is a relationship between the dc offset of a negative clamper and the *positive* peak value of the output waveform. Based on your results, state and explain this relationship.

4. Based on your answer to Question 3, what is the relationship between the dc offset of a positive clamper and the *negative* peak value of its output waveform?

5. Refer to the circuit shown in Figure 30.5. How would you decrease the ripple in the output of the voltage doubler? (*Hint:* Review Exercise 29, considering the *RC* time constants in the filtered rectifier.)

6. Refer to the V_{dc} value that you measured in Step 9. Was this value *exactly* twice the peak input voltage? If not, explain why.

7. Refer to the circuit shown in Figure 30.5. This is a *positive* voltage doubler. In the space below, sketch a *negative* half-wave voltage doubler.

SIMULATION EXERCISE

Procedure

1. Open file Ex30.1 from the Electronics Technology Fundamentals companion web site (www.prenhall.com/paynter). This file contains all three circuits covered in the hardware portion of this exercise. Run the simulation to make certain that all three circuits are functioning properly before introducing any faults.
2. Insert each of the faults that follow into the appropriate circuits. Run the simulation and record your results/analysis for each fault in the fault analysis charts that you copied from Appendix D.

> *Note:* Some faults that are inserted using the simulator can be dangerous to equipment and/or personnel when they occur in real circuits. Simulations provide exposure to such faults in a safe environment.

Fault Simulations

Fault 30-1	Open diode in the shunt clipper.
Fault 30-2	Open diode in the clamper.
Fault 30-3	Open shunt diode (D_2) in the voltage multiplier.
Fault 30-4	Open output capacitor (C_3) in the voltage multiplier.
Fault 30-5	Open output diode (D_3) in the voltage multiplier.

Exercise 31

Base Bias

OBJECTIVES

After completing this exercise, you should be able to:

- Calculate the *dc forward current gain* (h_{FE}) for a transistor using measured values of I_C and I_B.
- Demonstrate the relationship among I_B, I_C, and V_{CE} for a bipolar junction transistor (BJT).
- Draw the dc load line for a base-bias circuit.
- Use the dc load line to identify combinations of I_C and V_{CE} for a BJT circuit.
- Demonstrate the relative instability of the base-bias circuit.

DISCUSSION

Base bias is the simplest of the transistor biasing circuits. It consists of a single transistor, two resistors, and a single dc power supply. (An added potentiometer is used in this exercise to provide for circuit adjustments.) In the first part of this exercise, you will use a base-bias circuit to examine the relationships among I_B, I_C, and V_{CE}.

A *dc load line* is a graph that represents every possible combination of I_C and V_{CE} for a specific transistor-biasing circuit. In the second part of this exercise, you will generate the dc load line for the base-bias circuit shown in Figure 31.1.

It would seem that the simplicity of the base-bias circuit would make it ideal for most applications; however, the base-bias circuit is relatively unstable against changes in beta and/or temperature. This point will be examined in the third part of this exercise.

Review:

· Sections 19.1 through 19.5 of *Electronics Technology Fundamentals*.
· The spec sheet for the 2N3904 *npn* transistor.

MATERIALS

1	variable dc power supply
2	DMMs
1	soldering iron
1	protoboard
2	resistors: 2 kΩ and 100 kΩ
1	2 MΩ precision potentiometer
2	2N3904 *npn* transistors

PROCEDURE

Part 1: BJT Voltage and Current Characteristics

1. Measure and record the values of R_{B2} and R_C for the circuit shown in Figure 31.1.

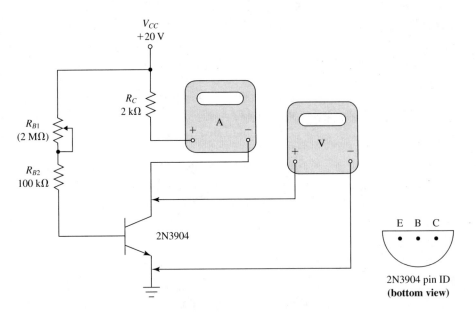

FIGURE 31.1 Base bias.

Component	Nominal Value	Measured Value
R_{B2}		
R_C		

2. Construct the circuit shown in Figure 31.1. Wait to connect the voltmeter until directed to do so.
3. Adjust R_{B1} until you get a reading of 2 mA in the collector circuit.
4. Measure V_{RB2} with your DMM, and use the measured value of R_{B2} to calculate I_B. Then, measure V_{CE} and record your results in Table 31.1.

TABLE 31.1   Current Values for the Transistor Shown in Figure 31.1

I_C	I_B	V_{CE}	h_{FE}
2 mA			
4 mA			
6 mA			
8 mA			

5. Repeat Steps 3 and 4 for I_C values of 4 mA, 6 mA, and 8 mA. Record your results in Table 31.1.
6. Use the values of I_C and I_B to calculate the dc forward current gain (h_{FE}) for the transistor at each value of I_C. Record these calculations in Table 31.1 as well.

Part 2: Base Bias

7. Using your measured value of R_C and a V_{CC} of 20 V, calculate $I_{C(sat)}$, and then plot the dc load line for this circuit in Figure 31.2.

$$I_{C(sat)} = \underline{\hspace{3cm}}$$

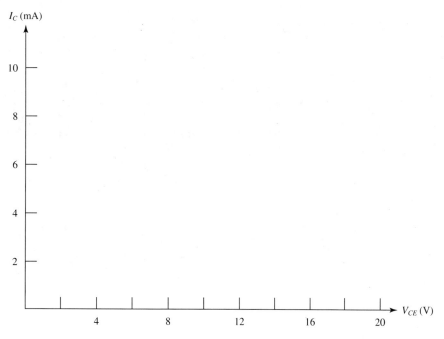

FIGURE 31.2   The dc load line for the circuit shown in Figure 31.1.

8. Use the load line shown in Figure 31.2 to predict the value of I_C for V_{CE} values of 2 V, 8 V, and 16 V. Enter these values in Table 31.2.

TABLE 31.2   Predicted and Measured Current Values

V_{CE}	I_C (Predicted)	I_C (Measured)
2 V		
8 V		
16 V		

9. Adjust R_{B1} until V_{CE} = 2 V. Measure I_C, and record this value in Table 31.2. Repeat for the other V_{CE} values listed in Step 8, and enter these values in the table.

Part 3: Base-Bias Instability

10. Power up, and adjust R_{B1} until the circuit is midpoint biased (V_{CE} is one-half of V_{CC}). For this value of V_{CE}, measure I_C, and enter your values in the first column of Table 31.3.

TABLE 31.3   Bias-Instability Test

Value	First Transistor	Second Transistor	Change
V_{CEQ}			
I_{CQ}			

11. Without disturbing R_{B1}, power down, and swap the 2N3904 transistors. Restore power, and measure V_{CE} and I_C for this new transistor *without adjusting the R_{B1} setting*. Then, calculate the change in these two values due to the change in transistors. Enter your values in the appropriate columns in Table 31.3.

12. Adjust R_{B1} until the circuit is again midpoint biased. Heat up your soldering iron. When it is hot, *very carefully* touch the emitter terminal of the transistor with your soldering iron. Hold it to the emitter terminal of the transistor only long enough for an obvious change in the meter readings to occur. Detail these changes in the space provided below.

QUESTIONS & PROBLEMS

1. Refer to your results in Table 31.1 and the spec sheet for the 2N3904 transistor. Compare the values of h_{FE} that you calculated with the range given on the spec sheet. Were the h_{FE} values within specifications?

2. Did h_{FE} change as I_C changed? Explain why you think this was (or was not) the case.

3. Refer to your results in Table 31.1. Based on these results, explain the relationship between I_C and V_{CE}.

4. Refer to your results in Table 31.1. First, you doubled the value of I_C. Then, you increased it by one-half, and then by one-third. Based on your results, calculate the percent of change in I_B and V_{CE} for these changes in I_C, and state the relationship among I_C, I_B, and V_{CE}.

5. Refer to your results in Table 31.2. Did the values for I_C that you determined from the load line agree with your measured values? If not, explain any discrepancies.

6. Refer to your results in Table 31.3. Which of the two transistors had the higher h_{FE}? Explain your reasoning.

7. Refer to your results from Step 10.
    a. What do these results tell you about the effect of temperature on base-bias stability?

    b. What do you think happened to the transistor's h_{FE} when the component was heated? Explain your reasoning.

Procedure

1. Open file Ex31.1 from the Electronics Technology Fundamentals companion web site (www.prenhall.com/paynter). This file contains the circuit shown in Figure 31.1, but with an added ammeter in the base circuit.
2. Run the simulation and adjust R_{B1} so that $I_C \cong 5$ mA. Record the values of I_C, I_B, and V_{CE} in Table 31.4. The values represent the *no-fault* condition for the circuit.

TABLE 31.4 Fault Analysis

Fault	I_C	I_B	V_{CE}
No fault			
R_{B2} open			
R_{B1} and R_{B2} shorted			
R_C open			
R_C shorted			
Collector-emitter shorted			
Base-emitter shorted			
Base-emitter open			

Fault Simulation

Note: Some faults that are inserted using the simulator can be dangerous to equipment and/or personnel when they occur in real circuits. Simulations provide exposure to such faults in a safe environment.

Insert each fault listed in Table 31.4, and run the simulation. Record the values of I_C, I_B, and V_{CE} in Table 31.4 for each of the fault conditions. When you change from one fault to the next, make certain that the previous faulted component has been returned to its no-fault condition.

Questions

1. Explain your readings with R_{B2} open.

2. Explain your readings with both base resistors shorted.

3. Explain your readings with R_C open.

4. Explain your readings with R_C shorted.

5. Explain your readings with the collector-emitter junction shorted.

6. Explain your readings with the base-emitter junction shorted.

7. Explain your readings with the base-emitter junction open.

Exercise 32

Voltage-Divider Bias

OBJECTIVES

After completing this exercise, you should be able to:

- Perform a dc analysis of a voltage-divider bias circuit.
- Explain the effect that transistor loading has on the base voltage of a voltage-divider bias circuit.
- Demonstrate the relative stability of the voltage-divider bias circuit.

DISCUSSION

Voltage-divider bias is the most commonly used type of BJT biasing for two primary reasons:

- It provides excellent operating point (Q) stability, even with wide variations in h_{FE} and temperature.
- It requires only one supply voltage. (Its closest competitor, in terms of bias stability, requires *two* supply voltages.)

    In the first part of this exercise, you will perform a basic dc analysis of a voltage-divider bias circuit. In the second part, you will look at how transistor loading affects V_B. Finally, in the third part, you will attempt to prove that voltage-divider bias has much better Q-point stability than base bias.

LAB PREPARATION

Review:

- Section 19.6 of *Electronics Technology Fundamentals*.
- The spec sheet for the 2N3904 transistor.
- Review the discussion on fault analysis in Appendix D.
- Make eight copies of the *fault analysis chart* in Appendix D (for the fault simulations portion of the exercise).

MATERIALS

1	variable dc power supply
1	DMM
1	soldering iron
1	protoboard
6	resistors: 560 Ω, 1.5 kΩ, 6.8 kΩ, 33 kΩ, 68 kΩ, and 330 kΩ
2	2N3904 *npn* transistors

PROCEDURE

Part 1: Voltage-Divider Bias dc Analysis

1. Measure and record the values of the resistors used in the circuit shown in Figure 32.1.

Component	Nominal Value	Measured Value
R_1		
R_2		
R_E		
R_C		

2. Using the measured values of the resistors, perform a dc analysis of the circuit shown in Figure 32.1. Record the results of your calculations below.

$$V_B = V_{CC}\frac{R_2}{R_1 + R_2} = \underline{\hspace{3cm}}$$

$$V_E = V_B - V_{BE} = \underline{\hspace{3cm}}$$

$$I_C \cong I_E = \frac{V_E}{R_E} = \underline{\hspace{3cm}}$$

$$V_{RC} = I_C R_C = \underline{\hspace{3cm}}$$

$$V_{CEQ} = V_{CC} - I_C(R_C + R_E) = \underline{\hspace{3cm}}$$

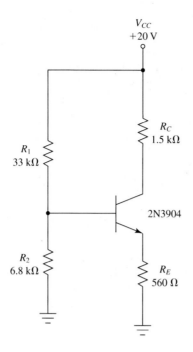

2N3904 pin ID
(**bottom view**)

FIGURE 32.1

3. Construct the circuit shown in Figure 32.1. Use your DMM to measure the following values:

$$V_{R1} = \underline{\hspace{2cm}}$$

$$V_B = \underline{\hspace{2cm}}$$

$$V_{RE} = \underline{\hspace{2cm}}$$

$$V_{RC} = \underline{\hspace{2cm}}$$

$$V_{CEQ} = \underline{\hspace{2cm}}$$

4. Based on your measurements recorded in Step 3, make the following calculations:

$$V_{BE} = V_B - V_E = \underline{\hspace{2cm}}$$

$$I_E = \frac{V_E}{R_E} = \underline{\hspace{2cm}}$$

$$I_C = \frac{V_{RC}}{R_C} = \underline{\hspace{2cm}}$$

$$I_{R1} = \frac{V_{R1}}{R_1} = \underline{\hspace{2cm}}$$

$$I_{R2} = \frac{V_{R2}}{R_2} = \underline{\hspace{2cm}}$$

$$I_B = I_{R1} - I_{R2} = \underline{\hspace{2cm}}$$

$$h_{FE} = \frac{I_C}{I_B} = \underline{\hspace{2cm}}$$

$$R_{base} = h_{FE}R_E = \underline{\hspace{2cm}}$$

5.   Modify the circuit as shown in Figure 32.2. The ratio of R_1 to R_2 is still the same, but the loading effect of the transistor has changed as a result of the modification. Measure the following values:

$V_{R1} = $ _____ $V_{R2} = $ _____

$V_B = $ _____ $V_{RE} = $ _____

$V_{RC} = $ _____ $V_{CEQ} = $ _____

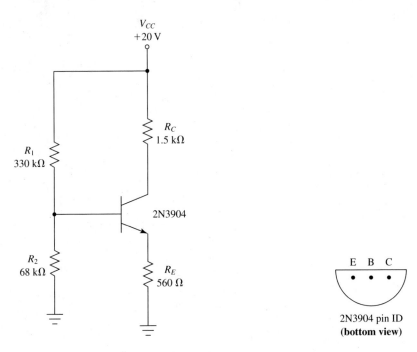

FIGURE 32.2   The modified voltage-divider bias circuit.

6.   Based on your measurements in Step 5, calculate the following values:

$$V_{BE} = V_B - V_E = \text{_____}$$

$$I_E = \frac{V_E}{R_E} = \text{_____}$$

$$I_{CQ} = \frac{V_{RC}}{R_C} = \text{_____}$$

$$I_{R1} = \frac{V_{R1}}{R_1} = \text{_____}$$

$$I_{R2} = \frac{V_{R2}}{R_2} = \text{_____}$$

$$I_B = I_{R1} - I_{R2} = \text{_____}$$

$$h_{FE} = \frac{I_C}{I_B} = \text{_____}$$

$$R_{\text{base}} = h_{FE}R_E = \text{_____}$$

Part 3: Voltage-Divider Bias Stability

7. Replace the base-biasing resistors with the original values shown in Figure 32.1. Swap the 2N3904 transistors. Measure V_{CE}, and record this value below. Then, measure V_{RC}, and use this value to calculate I_{CQ}. Record these values below as well.

$$V_{CEQ} = \underline{\hspace{2cm}}$$

$$V_{RC} = \underline{\hspace{2cm}}$$

$$I_{CQ} = \underline{\hspace{2cm}}$$

8. Heat your soldering iron, and connect the DMM to monitor V_{CEQ}. When the iron is hot, *very carefully* touch it to the emitter terminal of the 2N3904. Leave the iron in contact with the transistor for 1 or 2 seconds. Record any changes in V_{CE} below.

$$\Delta V_{CE} = \underline{\hspace{2cm}} \text{ V}$$

QUESTIONS & PROBLEMS

1. Refer to your results in Steps 2 through 4. How close were your calculated values to measured values? Explain any discrepancies.

2. Refer to the V_B values that you calculated in Step 2, and measured in Steps 3 and 5. In which circuit was transistor loading more pronounced? Explain your answer.

3. Refer to your calculated values of h_{FE} in Steps 4 and 6.

   a. Did these values fall within the range listed on the 2N3904 spec sheet?

   b. Did the h_{FE} value change for the two circuits? If so, explain why you think there was a difference.

4. Your results in Part 2 of this exercise demonstrated that using high-value bias resistors results in the emitter circuit loading down R_2. It would seem logical to simply use low-value bias resistors to avoid this problem. Why do you think this is not always done in practice? (*Hint:* Think about power dissipation and efficiency.)

5. Refer to Steps 7 and 8. What do your results tell you about the Q-point stability of the voltage-divider bias circuit when compared to the base-bias circuit that you studied in Exercise 31? Explain your answer.

Discussion

One of the strengths of a simulator is that it allows you to perform measurements that would be impractical or impossible in a real-life situation. Normally, a technician must troubleshoot a circuit using only a voltmeter and perhaps an oscilloscope. It is usually impractical to use ammeters as the circuit must be desoldered and the ammeter installed. To use 5 ammeters and 2 voltmeters simultaneously, as in this simulation, would never happen in real life.

The reason we are using so many test instruments is to allow you to see what is happening throughout the circuit. This way you can see the interdependency of the base, emitter, and collector circuits.

Procedure

1. Open file Ex32.1 from the Electronics Technology Fundamentals companion web site (www.prenhall.com/paynter).
2. Run the simulation and record the readings of the seven meters in Table 32.1. These readings are the *no-fault* values for the circuit.

TABLE 32.1 Simulation Results

Current	Measured Value	Voltage	Measured Value
I_{R1}		V_B	
I_{R2}		V_C	
I_B		V_E	
I_E			
I_C			

3. Run each of the fault simulations listed below. Record your results/analysis for each fault in one of the fault analysis charts. Remember to return each component to its no-fault condition before introducing the next fault.

Fault Simulations

> *Note:* Some faults that are inserted using the simulator can be dangerous to equipment and/or personnel when they occur in real circuits. Simulations provide exposure to such faults in a safe environment.

Fault Simulations

Fault 32-1   Open bias resistor (R_1).

Fault 32-2   Open transistor base lead.

Fault 32-3   Open transistor emitter lead.

Fault 32-4   Open transistor collector lead.

Fault 32-5   Shorted transistor (collector-to-emitter).

Fault 32-6   Shorted transistor (base-to-emitter).

Fault 32-7   Shorted collector resistor

Fault 32-8   Open bias resistor (R_2).

Questions

1. Compare the values of I_{R1} and I_{R2} in Table 32.1. Are they nearly equal in value? Should they be? Why or why not?

2. Compare the values of I_C and I_E in Table 32.1. Are they nearly equal in value? Should they be? Why or why not?

3. Do the transistor terminal currents comply with Kirchhoff's current law? Explain your answer.

4. Using the values in Table 32.1 and resistor values from the circuit, demonstrate that the emitter–collector–power supply loop of the circuit complies with Kirchhoff's voltage law.

Exercise 33

The Common-Emitter Amplifier

OBJECTIVES

After completing this exercise, you should be able to:

- Determine the voltage gain of a common-emitter (CE) amplifier.
- Determine the input impedance of a CE amplifier.
- Determine r_e' for a CE amplifier.
- Explain the role of coupling and bypass capacitors in a CE amplifier.
- Demonstrate the effect that a load has on the voltage gain of a CE amplifier.

DISCUSSION

The *common-emitter (CE) amplifier* is the most common BJT amplifier configuration. This amplifier has moderate voltage and current gain, which means it has relatively high power gain. It is the only BJT amplifier with a 180° voltage phase shift between its input and output waveforms.

In this exercise, you will examine most of the important characteristics of a CE amplifier. You will calculate and measure the amplifier's voltage gain and input impedance, and you will examine the effects of bypass capacitors and load resistance on the voltage gain of the amplifier.

· Review Sections 20.1 through 20.4 of *Electronics Technology Fundamentals*.

· Review the discussion on fault analysis in Appendix D.

· Make nine copies of the *fault analysis chart* in Appendix D (for the fault simulations portion of the exercise).

MATERIALS

1	variable dc power supply
1	dual-trace oscilloscope
1	protoboard
6	resistors: 390 Ω, 1 kΩ, 1.5 kΩ, 2.2 kΩ, 10 kΩ, and 12 kΩ
3	capacitors: 10 μF (2) and 100 μF
1	2 kΩ precision potentiometer
1	2N3904 *npn* transistor

PROCEDURE

1. Calculate the following values for the circuit shown in Figure 33.1 (assume that $h_{FE} = h_{fe} = 200$):

$$V_B = \underline{\hspace{3cm}}$$

$$V_E = \underline{\hspace{3cm}}$$

$$I_E = \underline{\hspace{3cm}}$$

$$r_e' \cong \frac{25 \text{ mV}}{I_E} = \underline{\hspace{3cm}}$$

$$r_C = \underline{\hspace{3cm}}$$

$$A_v = \frac{r_C}{r_e'} = \underline{\hspace{3cm}}$$

$$Z_{in} = R_1 \parallel R_2 \parallel h_{fe}r_e' = \underline{\hspace{3cm}}$$

2. Construct the circuit shown in Figure 33.1. Channel 1 (C1) is connected to display the input signal. Channel 2 (C2) is connected to display the output (load) signal. Both channels should be ac coupled.

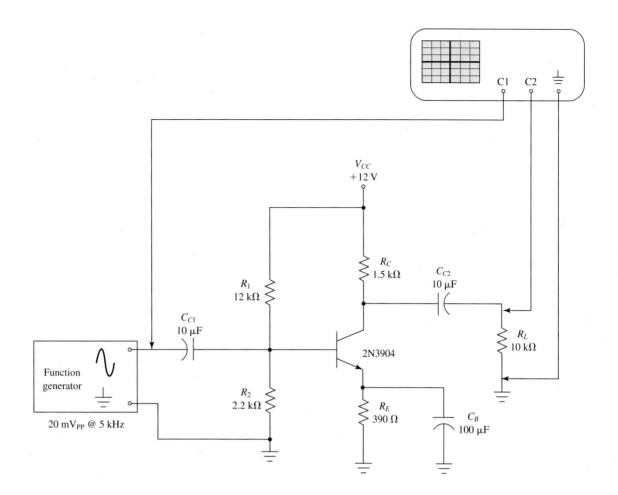

FIGURE 33.1 Common-emitter (CE) amplifier.

3. Adjust the input signal to the value shown in Figure 33.1. Measure the peak-to-peak values of the input and output waveforms, and record these values below. Use the measured signal voltages to calculate the voltage gain of the amplifier.

$$v_{in} = \underline{\hspace{3cm}}$$

$$v_{out} = \underline{\hspace{3cm}}$$

$$A_v = \frac{v_{out}}{v_{in}} = \underline{\hspace{3cm}}$$

4. Carefully draw the input and output waveforms in Figure 33.2.

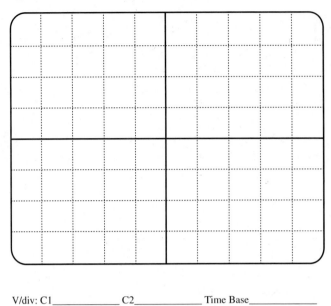

V/div: C1_____ C2_____ Time Base_____

FIGURE 33.2   CE amplifier input and output waveforms.

5. Using the basic laws of voltage division, you can easily determine the input impedance of this amplifier as follows:

    • Insert the 2 kΩ potentiometer between the signal generator and the input coupling capacitor, as shown in Figure 33.3.

    • Connect Channel 1 to the input side of the 2 kΩ pot. Connect Channel 2 to the other side of the pot.

    • Adjust the potentiometer until the signal on the transistor side of the pot is exactly one-half the input voltage.

    • Power down, and remove the pot from the circuit *without disturbing its setting*.

    • Measure the adjusted resistance of the potentiometer, and record this value below. This value equals the input impedance of the amplifier.

$$Z_{in} = \text{\underline{\hspace{3cm}}}$$

6. Restore the circuit to its original configuration shown in Figure 33.1. Change R_L to 1 kΩ, and repeat Step 3. Record your measured and calculated values below.

$$v_{in} = \text{\underline{\hspace{3cm}}}$$

$$v_{out} = \text{\underline{\hspace{3cm}}}$$

$$A_v = \frac{v_{out}}{v_{in}} = \text{\underline{\hspace{3cm}}}$$

7. Restore the load to 10 kΩ, and remove the bypass capacitor (C_B). Repeat Step 3, and record your measured and calculated values below.

$$v_{in} = \text{\underline{\hspace{3cm}}}$$

$$v_{out} = \text{\underline{\hspace{3cm}}}$$

$$A_v = \frac{v_{out}}{v_{in}} = \text{\underline{\hspace{3cm}}}$$

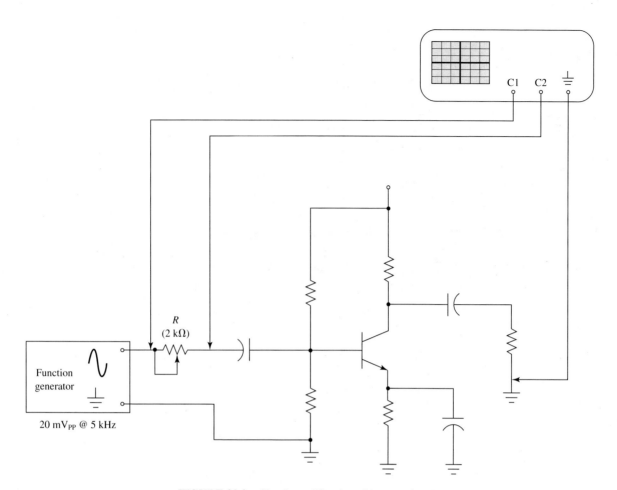

FIGURE 33.3   Circuit modifications for measuring Z_{in}.

QUESTIONS & PROBLEMS

1. Refer to Steps 1 and 3. How close was your calculated value of A_v to the measured value? Explain any discrepancies.

2. Is the amplifier shown in Figure 33.1 midpoint biased? Support your answer using values from Step 1.

3. Refer to the waveforms that you drew in Figure 33.2. Explain why the output signal is 180° out of phase with the input.

4. Refer to Steps 1 and 5. How close was your calculated value of Z_{in} to the measured value? Explain any discrepancies.

5. Refer to your A_v calculation from Step 3. Using your measured value of A_v and the calculated value of r_C, solve for r_e'. Compare this value to your calculated value in Step 1. Explain any discrepancies.

$$r_e' = \frac{r_C}{A_v} = \underline{\hspace{3cm}}$$

6. Refer to Step 6. What happened to the voltage gain of the circuit when the load resistance was decreased? Explain why you think this happened.

7. Refer to Step 7. What happened to the voltage gain of the circuit when the bypass capacitor was removed? Explain why you think this happened.

SIMULATION EXERCISE

Procedure

1. Open file Ex33.1 from the Electronics Technology Fundamentals companion web site (www.prenhall.com/paynter).
2. Run the simulation and record the measurements listed below. Compare these values to those in the hardware portion of the lab to make sure the circuit is operating properly.

$$V_B = \text{_____} \qquad v_{in} = \text{_____}$$

$$V_E = \text{_____} \qquad v_{out} = \text{_____}$$

$$V_{CE} = \text{_____} \qquad A_v = \text{_____}$$

3. Run each of the fault simulations listed below. Record your results/analysis for each fault in one of the fault analysis charts. Remember to return each component to its no-fault condition before introducing the next fault.

Fault Simulations

> *Note:* Some faults that are inserted using the simulator can be dangerous to equipment and/or personnel when they occur in real circuits. Simulations provide exposure to such faults in a safe environment.

Fault Simulations

Fault 33-1   Open input coupling capacitor.

Fault 33-2   Open bias resistor (R_1).

Fault 33-3   Open transistor emitter lead.

Fault 33-4   Open transistor collector lead.

Fault 33-5   Shorted transistor (collector-to-emitter).

Fault 33-6   Open emitter bypass capacitor

Fault 33-7   Shorted transistor (base-to-emitter).

Fault 33-8   Open load

Fault 33-9   Open bias resistor (R_2).

Exercise 34

The Common-Collector Amplifier (Emitter Follower)

OBJECTIVES

After completing this exercise, you should be able to:

- Determine the voltage gain of a common-collector (CC) amplifier.
- Determine the input impedance of a CC amplifier.
- Demonstrate the effect that load resistance has on the input impedance and voltage gain of a CC amplifier.

DISCUSSION

The *common-collector (CC) amplifier* is the second most common BJT amplifier configuration. This amplifier has relatively high current gain and a voltage gain that is less than *unity* (1).

One common application of the CC amplifier is as a *buffer*. A buffer is used to compensate for an impedance mismatch between a load and a source. Because the emitter follower has relatively high input impedance and low output impedance, it is commonly used to couple a high-impedance source to a low-impedance load. You will look at this application in this exercise.

LAB PREPARATION

- Review Section 20.5 of *Electronics Technology Fundamentals*.
- Review the discussion on fault analysis in Appendix D.
- Make eight copies of the *fault analysis chart* in Appendix D (for the fault simulations portion of the exercise).

277

1 DMM
1 function generator
1 variable dc power supply
1 dual-trace oscilloscope
1 protoboard
5 resistors: 100 Ω, 1 kΩ (2), 16 kΩ, and 22 kΩ
1 10 kΩ precision potentiometer
3 capacitors: 0.1 μF, 10 μF, and 100 μF
1 2N3904 *npn* transistor

PROCEDURE

1. Refer to the circuit shown in Figure 34.1. Using the nominal resistor values, perform a dc analysis of the circuit. Record the following calculated values for this amplifier (assume that $h_{FC} = h_{fc} = 160$):

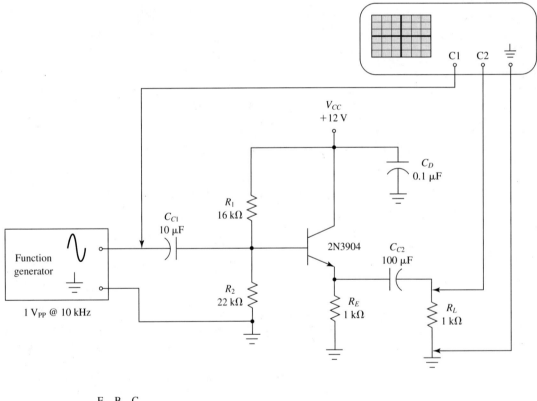

Note: C_D is a *decoupling capacitor*. This component is used to prevent a type of low-amplitude distortion that is common to this type of amplifier.

FIGURE 34.1   An emitter follower.

$$V_B = \rule{3cm}{0.4pt}$$

$$V_E = \rule{3cm}{0.4pt}$$

$$I_E = \rule{3cm}{0.4pt}$$

$$r_e' \cong \frac{25\ mV}{I_E} = \rule{3cm}{0.4pt}$$

$$r_E = R_E \parallel R_L = \rule{3cm}{0.4pt}$$

$$A_v = \frac{r_E}{r_e' + r_E} = \rule{3cm}{0.4pt}$$

$$Z_{in} = R_1 \parallel R_2 \parallel h_{fc}(r_e' + r_E) = \rule{3cm}{0.4pt}$$

2. Construct the circuit shown in Figure 34.1. Channel 1 (C1) of the oscilloscope is connected to display the input signal; Channel 2 (C2) is connected to display the output (load) signal. Both channels should be ac coupled.

3. Measure the input and output signal voltages and record their values in the table below. Then, use these voltages to calculate the voltage gain of the amplifier and record your result in the table as well.

4. Change the load to 100 Ω, and repeat the measurements and calculations in Step 3.

R_L	v_{in} (Measured)	v_{out} (Measured)	A_v (Calculated)
1 kΩ			
100 Ω			

5. Restore the load to its original value, and determine the input impedance of the amplifier as follows:

- Insert the 10 kΩ potentiometer between the signal generator and the input coupling capacitor as shown in Figure 34.2.

- Connect the oscilloscope as shown in Figure 34.2.

- Adjust the potentiometer until the signal on the capacitor side of the pot is exactly one-half the input voltage.

- Power down, and carefully remove the pot from the circuit *without disturbing its setting*.

Measure the resistance of the potentiometer, and record this value below. This value represents the input impedance of the amplifier.

$$Z_{in} = \rule{3cm}{0.4pt}$$

6. Change the load to 100 Ω and measure the amplifier input impedance.

$$Z_{in} = \rule{3cm}{0.4pt}$$

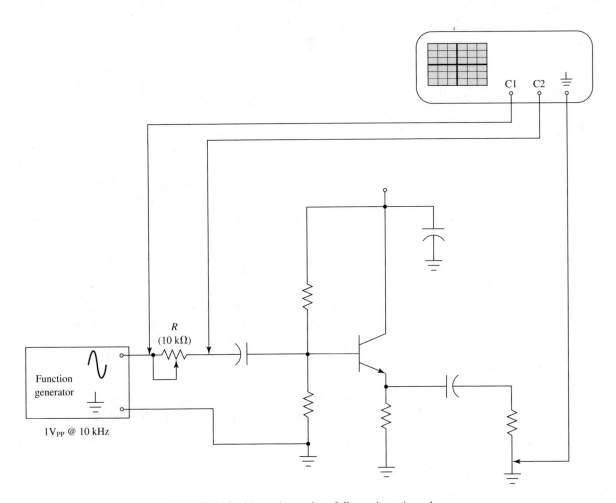

FIGURE 34.2 Measuring emitter-follower input impedance.

At the start of this exercise, it was stated that the CC amplifier is often used to couple a high-impedance source to a low-impedance load. In the following steps, you will see how effective the CC amp is in this application.

7. Adjust the 10 kΩ potentiometer installed in Step 5 to 1 kΩ. This will represent the output impedance of the source. Leave the 100 Ω load connected.
8. Adjust V_S for 1 V_{PP}, measured on the *output* side of the potentiometer.
9. Calculate load power as if the 100 Ω load was connected directly to the 1 kΩ source resistor, as follows:

$$V_{L(\text{pk})} = V_{S(\text{pk})} = \frac{R_L}{R_S + R_L} = \underline{\hspace{2cm}}$$

$$V_L = 0.707 V_{L(\text{pk})} = \underline{\hspace{2cm}}$$

$$P_L = \frac{V_L^2}{R_L} = \underline{\hspace{2cm}}$$

10. Measure the peak value of V_L, and convert this measurement to an rms value. Then calculate the load power. Record these values below.

$$V_{L(pk)} = \underline{\hspace{3cm}}$$

$$V_L = 0.707 V_{L(pk)} = \underline{\hspace{3cm}}$$

$$P_L = \frac{V_L^2}{R_L} = \underline{\hspace{3cm}}$$

QUESTIONS & PROBLEMS

1. Refer to your results in Steps 1 and 3. How close was your calculated value of A_v to the measured value? Explain any discrepancies.

2. Refer to your calculations in Step 1. Is this amplifier midpoint biased? Explain your answer.

3. Refer to your results from Steps 3 and 4. What effect did the change in load have on the voltage gain of the amplifier? Explain the change (if any).

4. Refer to your results in Steps 1 and 5. How close was your calculated value of Z_{in} to the measured value? Explain any discrepancies.

5. Refer to your results from Steps 5 and 6. What effect did the change in load have on the input impedance of the amplifier? Explain the change (if any).

6. Refer to your results in Steps 7 and 8. Was the amplifier effective in coupling the high-impedance source to the low-impedance load? Explain your answer.

SIMULATION EXERCISE

Procedure

1. Open file Ex34.1 from the Electronics Technology Fundamentals companion web site (www.prenhall.com/paynter).
2. Run the simulation and record the measurements listed below. Compare these values to those in the hardware portion of the lab to make sure the circuit is operating properly.

$V_B =$ _____ $v_{in} =$ _____

$V_E =$ _____ $v_{out} =$ _____

$V_{CE} =$ _____ $A_v =$ _____

3. Run each of the fault simulations listed below. Record your results/analysis for each fault in one of the fault analysis charts. Remember to return each component to its no-fault condition before introducing the next fault.

Fault Simulations

> *Note:* Some faults that are inserted using the simulator can be dangerous to equipment and/or personnel when they occur in real circuits. Simulations provide exposure to such faults in a safe environment.

Fault Simulations

Fault 34-1   Open input coupling capacitor.

Fault 34-2   Open bias resistor (R_1).

Fault 34-3   Open transistor emitter lead.

Fault 34-4   Shorted transistor (collector-to-emitter).

Fault 34-5   Open output coupling capacitor.

Fault 34-6   Shorted transistor (base-to-emitter).

Fault 34-7   Open load

Fault 34-8   Open bias resistor (R_2)

Exercise 35

Class B and Class AB Amplifiers

OBJECTIVES

After completing this exercise, you should be able to:

- Explain the source of crossover distortion in class B amplifiers.
- Demonstrate the reasons that class AB amplifiers are preferred over class B amplifiers.
- Determine load power in a class AB amplifier.
- Determine the efficiency of a class AB amplifier.

DISCUSSION

Both class B and class AB amplifiers employ two transistors: an *npn* for the positive half of the waveform, and a *pnp* for the negative half. The primary difference between class B and class AB amplifiers is the manner in which the transistors are biased. In the class B amplifier, each transistor is biased at cutoff. The class AB amplifier, however, uses two *compensating diodes* to bias the transistors just *above* cutoff. Both amplifiers are relatively efficient, because I_{CQ} is either very low or near zero. Both amplifiers also have the classic emitter-follower characteristics of relatively high current gain and a voltage gain of less than unity. In this exercise, you will examine these two designs and compare their attributes.

LAB PREPARATION

Review Section 20.6 of *Electronics Technology Fundamentals*.

1   function generator
1   variable dc power supply
1   dual-trace oscilloscope
1   DMM
1   protoboard
4   resistors: 27 Ω, 200 Ω, and 2.2 kΩ (2)
4   capacitors: 0.1 μF, 10 μF (2), and 470 μF
1   2N3904 *npn* transistor
1   2N3906 *pnp* transistor
2   1N4148 small-signal diodes

PROCEDURE

1. Construct the circuit shown in Figure 35.1. Before applying the input signal, use your DMM to measure the values listed in the table below. Record your measured values in the table.

Label	Measured Value
$V_{BE(Q1)}$	
$V_{BE(Q2)}$	
$V_{CE(Q1)}$	
$V_{CE(Q2)}$	

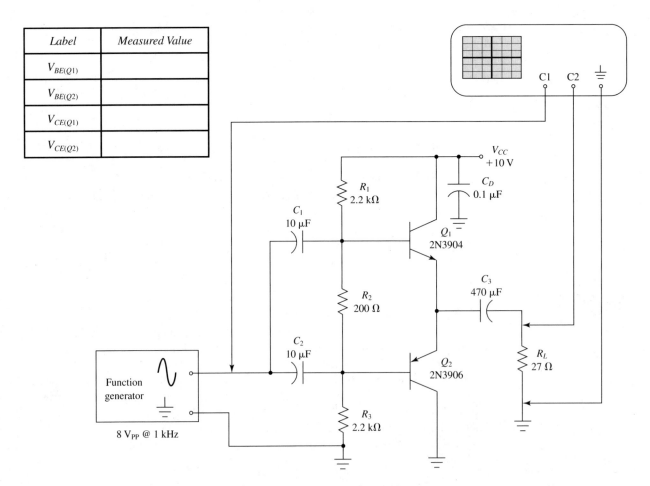

FIGURE 35.1   A class B amplifier.

2. Apply the input signal shown in Figure 35.1. Observe the input (Channel 1) and output (Channel 2) waveforms for the circuit. Draw both waveforms in Figure 35.2a.

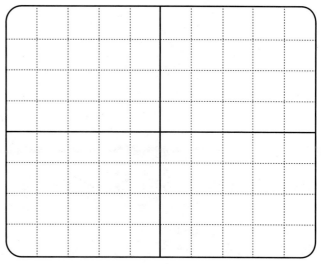

(a) Input and output waveforms for the circuit in Figure 35.1

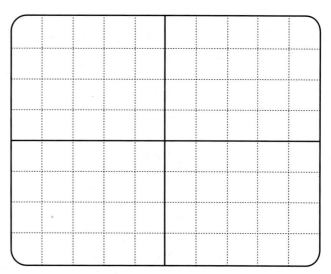

(b) Input and output waveforms for the circuit in Figure 35.3

FIGURE 35.2 Amplifier waveforms.

3. Replace R_2 with two 1N4148 diodes, as shown in Figure 35.3. Before applying the input signal, measure and record the values in the table below.

Label	Measured Value
$V_{BE(Q1)}$	
$V_{BE(Q2)}$	
$V_{CE(Q1)}$	
$V_{CE(Q2)}$	

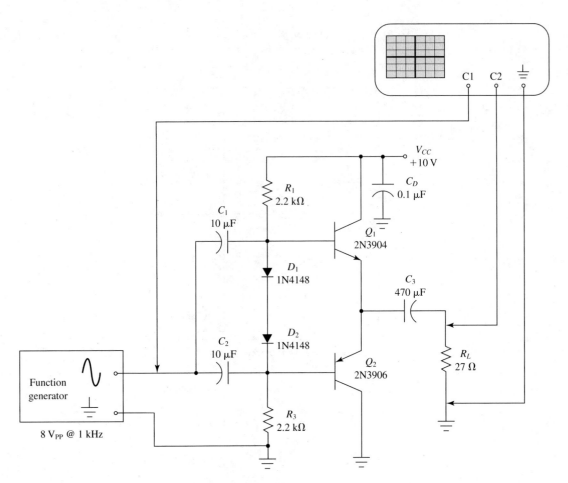

FIGURE 35.3   A class AB amplifier.

4. Apply the input signal shown. Observe the input (Channel 1) and output (Channel 2) waveforms for the circuit. Record the peak-to-peak input and output values, and draw both waveforms in Figure 35.2b.

$$v_{\text{in}} = \underline{\hspace{2cm}} \text{V}_{\text{pp}} \qquad v_{out} = \underline{\hspace{2cm}} \text{V}_{\text{pp}}$$

5. Use your results from Step 4 to calculate the voltage gain of the amplifier and the load power. (Remember to convert to rms values before you calculate load power.)

$$A_v = \underline{\hspace{2cm}}$$

$$P_L = \frac{V_L^2}{R_L} = \underline{\hspace{2cm}}$$

6. Insert a dc ammeter between the circuit and the dc supply. This will allow you to measure the total current drawn by the circuit (I_{CC}) from the dc supply. Measure and record the value of I_{CC}, then use this value and V_{CC} to calculate the source power (P_S).

$$I_{CC} = \underline{\hspace{2cm}}$$

$$P_S = I_{CC}V_{CC} = \underline{\hspace{2cm}}$$

7. Using the values calculated in Steps 5 and 6, calculate the efficiency of this amplifier.

$$\eta = \frac{P_L}{P_S} \times 100 = \underline{\hspace{2cm}}$$

QUESTIONS & PROBLEMS

1. Refer to Steps 1 and 3. Which amplifier had the highest values of V_{BE}? Explain why this was the case.

2. Refer to your A_v calculation in Step 5. Which type of amplifier configuration is this consistent with?

3. Refer to Step 3. Is this amplifier midpoint biased? If not, explain what component value(s) you would change to make it midpoint biased.

4. Refer to the circuit shown in Figure 35.3. Assume that R_1 and R_3 were changed to 10 kΩ. What effect would this have on the voltage gain and the input impedance of the circuit?

5. Refer to your efficiency calculations in Step 7. How did this value compare to the theoretical maximum efficiency for a class AB amplifier?

SIMULATION EXERCISE

Procedure

1. Open file Ex35.1 from the Electronics Technology Fundamentals companion web site (www.prenhall.com/paynter) and run the simulation. To confirm that the circuit is working properly, measure the input and output voltage, and then calculate the voltage gain of the amplifier. Record these values in Table 35.1.

TABLE 35.1   Class AB Amplifier Simulation Results

v_{in} *(Measured)*	v_{out} *(Measured)*	A_v *(Calculated)*

2. In Question 4 of the hardware portion of this exercise, you were asked what effect increasing the bias resistors would have on the circuit. Modify the circuit so that $R_1 = R_2 = 10$ kΩ. Run the simulation, and draw the output waveform in Figure 35.4.

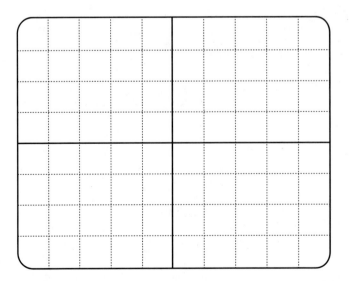

FIGURE 35.4

3. There are two ac ammeters and one dc ammeter provided on your workspace. Insert one ac ammeter (labeled as input current) between the ac source and input coupling capacitors. Insert the other ac ammeter (labeled as output current) between the output coupling capacitor and the load. Run the simulation, measure the input and output ac currents, and record these values in the first row of Table 35.2.

4. Change R_1 and R_2, first to 2.2 kΩ and then to 1 kΩ, and repeat Step 3.

TABLE 35.2 Class AB Amplifier Current Mesurements

$R_1 = R_2$	i_{in} (Measured)	i_{out} (Measured)	A_i (Calculated)
10 kΩ			
2.2 kΩ			
1 kΩ			

5. Return the circuit to its original values. Insert the dc ammeter between the dc supply and the circuit to measure I_{CC}. Run the simulation, and record the value for I_{CC} below. Then, use this value and V_{CC} to calculate P_S.

$$I_{CC} = \underline{\hspace{3cm}}$$

$$P_S = \underline{\hspace{3cm}}$$

6. Fault D_1 open, and repeat Step 5. Record your measurements below.

$$I_{CC} = \underline{\hspace{3cm}}$$

$$P_S = \underline{\hspace{3cm}}$$

7. Use the oscilloscope to monitor the input and output waveforms. Describe the output waveform. Is there any crossover distortion?

Questions

1. Refer to your waveform in Figure 35.4. Explain why the amplifier is acting like a class B amplifier.

2. Refer to your results in Table 35.2. What happened to the current gain of the circuit when R_1 and R_2 were decreased? Explain why you think this happened.

3. Refer to your results from Steps 5 through 7 of the simulation. How would you explain the lack of crossover distortion?

Exercise 36

JFET Operation

OBJECTIVES

After completing this exercise, you should be able to:

- Demonstrate the effect of drain-source voltage (V_{DS}) on drain current (I_D).
- Demonstrate the effect of gate-source voltage (V_{GS}) on drain current (I_D).
- Measure the *pinch-off voltage* (V_P) and *shorted-gate drain current* (I_{DSS}) for a junction field-effect transistor (JFET).
- Plot a JFET drain curve, and identify the *ohmic* and *constant-current* regions of the curve.
- Plot the transconductance curve for a given JFET.

DISCUSSION

Unlike the BJT, the JFET is a *voltage-controlled* device. In the first part of this exercise, you will examine the effect of drain-source voltage (V_{DS}) on drain current (I_D). You will measure I_D at various values of V_{DS}, and you will use these measurements to plot a JFET drain current curve.

In the second part of the exercise, you will examine the effect of gate-source voltage (V_{GS}) on drain current. You will measure I_D at various values of V_{GS}, and you will use these results to plot the *transconductance curve* for the device.

293

Review:

- Section 21.1 of *Electronics Technology Fundamentals*.
- The specification sheet for the 2N5486 *n*-channel JFET.

MATERIALS

1   variable dc power supply
2   DMMs
1   protoboard
3   resistors: 100 Ω (2) and 1 kΩ
1   10 kΩ potentiometer
1   2N5486 *n*-channel JFET

PROCEDURE

Part 1: Drain Current and V_{DS}

1. Refer to the specification sheet for the 2N5486 *n*-channel JFET. Obtain the rated ranges of V_P and I_{DSS} for the device. List these values below.

    $$V_P = \underline{\hspace{2cm}} \text{ to } \underline{\hspace{2cm}}$$

    $$I_{DSS} = \underline{\hspace{2cm}} \text{ to } \underline{\hspace{2cm}}$$

    > *Note:* V_P is the positive equivalent of the $V_{GS(\text{off})}$ rating.

2. Construct the circuit shown in Figure 36.1. Note that both the gate and source of the device are tied to ground. As a result, $V_{GS} = 0$ V.

    > *Note:* The meters are drawn as symbols to simplify Figure 36.1. Be sure to observe the proper polarity when inserting the meters in the circuit.

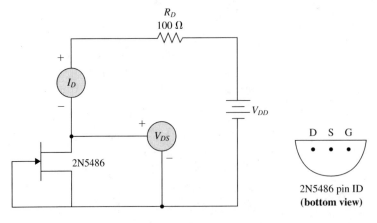

FIGURE 36.1   JFET test circuit.

3. Measure I_D with V_{DS} set to 0 V. Record this value in Table 36.1. Repeat this measurement for the remaining values of V_{DS} listed in the table.

TABLE 36.1   Voltage and Current Values for the JFET Test Circuit

V_{DS} (V)	I_D (mA)	V_{DS} (V)	I_D (mA)
0.0		4.0	
0.5		4.5	
1.0		5.0	
1.5		6.0	
2.0		7.0	
2.5		8.0	
3.0		10.0	
3.5		15.0	

4. Use the points from Table 36.1 to plot a curve on the graph in Figure 36.2. (This curve should be similar to the one shown in Figure 21.6a of *Electronics Technology Fundamentals*.)

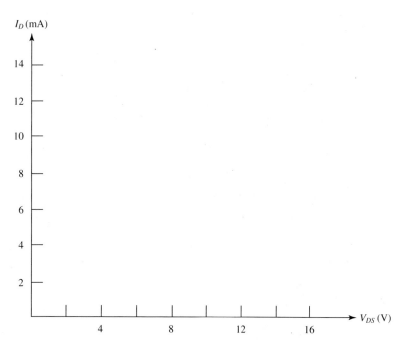

FIGURE 36.2   JFET drain curve.

5. Use your curve to determine V_P and I_{DSS} for this JFET, and record these values below.

$$V_P = \underline{\hspace{2cm}} \qquad I_{DSS} = \underline{\hspace{2cm}}$$

Note: V_P and I_{DSS} are measured at the point where the drain curve enters the *constant-current region*, as shown in Figure 21.6a in *Electronics Technology Fundamentals*.

The term *constant* may require some definition or explanation, since it is a relative term, relating the change in current per unit change in voltage. If a change of 1 V causes a change of 1 mA, it is obvious that the current is changing. If the same change in voltage causes a 0.1 mA (100 μA) change in current, then the rate of change has decreased. When a 1 V change causes only a 10 μA or 1 μA change in current, the rate of change has decreased to the point where we refer to the current as constant. (An argument could be made that *constant* refers to the slope of the curve, but this cannot be seen until after the data has been collected and the curve has been plotted.)

Part 2: JFET Transconductance

6. Construct the circuit shown in Figure 36.3. V_{DD} should initially be set to 0 V. R_{G2} should be adjusted so that V_{GS} is also at 0 V. Slowly increase V_{DD} while monitoring I_D. When I_D just starts to levels off, you have found I_{DSS}. Record this value below.

$$I_{DSS} = \text{_____}$$

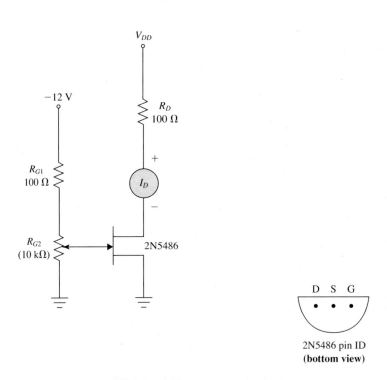

FIGURE 36.3   JFET test circuit.

7. Leave V_{DD} constant and adjust R_{G2} until I_D reaches its lowest level ($I_D \cong 0$ mA). Record the value of V_{GS} at which this occurs.

$$V_{GS(\text{off})} = \text{_____}$$

8. Plot the points representing I_{DSS} and $V_{GS(\text{off})}$ in Figure 36.4.

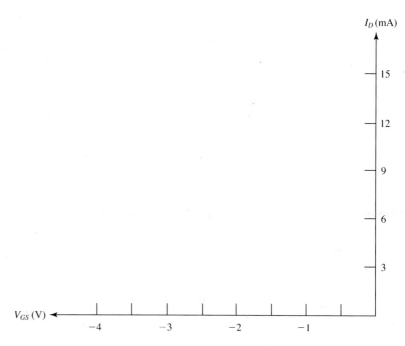

FIGURE 36.4   JFET transconductance curve.

9. Select four values of V_{GS} that fall between 0 V and $V_{GS(\text{off})}$. Record these values in Table 36.2. (If possible, use values that correspond to the increments on your graph.)

TABLE 36.2   Measured Combinations of V_{GS} and I_D

V_{GS}	I_D

10. Adjust R_{G2} so that V_{GS} equals the first value in Table 36.2. Measure I_D for this value of V_{GS}, and record the value in the table. Repeat this step for each value of V_{GS} listed in the table.

11. Use the values in Table 36.2 to plot the *transconductance curve* of this JFET in Figure 36.4. This curve should resemble the one shown in Figure 21.8 of *Electronics Technology Fundamentals*.

1. Refer to the values in Steps 1 and 5. How do your values of V_P and I_{DSS} compare to the rated values for the 2N5486 n-channel JFET?

2. Refer to Figure 36.2. Based on the curve, what is the relationship between V_{DS} and I_D when V_{DS} is less than V_P? What is this region of the graph called?

3. Refer to Figure 36.4. For each value of V_{GS}, calculate the associated value of I_D using

$$I_D = I_{DSS}\left[1 - \frac{V_{GS}}{V_{GS(\text{off})}} \right]^2$$

Record these values in Table 36.3. Also record your measured values from Table 36.2, and calculate the percent of variation between these two values.

TABLE 36.3 Calculated and Measured Drain Current Values

V_{GS}	I_D (Calculated)	I_D (Measured)	% Variation

How would you explain the variations between your measured and your calculated values in Table 36.3?

Discussion

This exercise requires that a variable dc source be used while running the simulation. While the source voltage can be varied, it requires restarting the simulation for each change. Using a fixed source along with a potentiometer provides a very useful, adjustable source for simulation circuits.

In this simulation exercise, you will plot the transconductance curve for the 2N5486. Using this JFET, you will plot a more exact curve using more combinations of I_D and V_{GS}.

Procedure

1. Open file Ex36.1 from the Electronics Technology Fundamentals companion web site (www.prenhall.com/paynter). This circuit is very similar to the test circuit shown in Figure 36.3.

2. Adjust the value of the potentiometer so that $V_{GS} = 0$ V, and make certain that V_{DS} is very close to 15 V. Note that this is the V_{DS} value at which I_{DSS} is determined as per the specification sheet. Change the value of V_{DD} if necessary. Measure I_{DSS} for $V_{DS} = 15$ V, and record this value below.

$$I_{DSS} = \underline{\hspace{3cm}}$$

3. Now, increase V_{GS} until I_D stops decreasing (it will be close to 0 mA), and determine $V_{GS(\text{off})}$. Record this value below.

$$V_{GS(\text{off})} = \underline{\hspace{3cm}}$$

4. Choose six approximately equidistant values of V_{GS} between 0 V and $V_{GS(\text{off})}$. Record these values in Table 36.4.

TABLE 36.4   Measured Combinations of V_{GS} and I_D

V_{GS}	I_D

5. For the values of V_{GS} listed in Table 36.4, measure the corresponding values of I_D. Record these values in the table.

6. Use the values of $V_{GS(off)}$, I_{DSS}, and those in Table 36.4 to plot the transconductance curve in Figure 36.5.

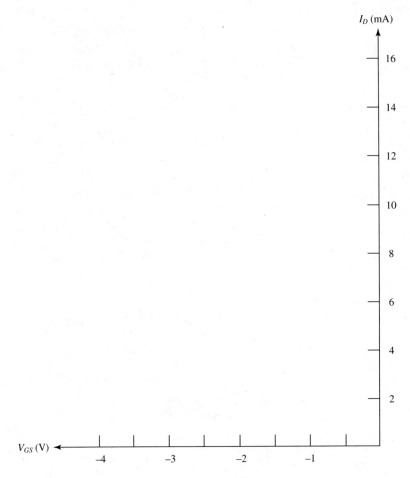

FIGURE 36.5 Simulation circuit transconductance curve.

Exercise 37

Gate-Bias and Self-Bias Circuits

OBJECTIVES

After completing this exercise, you should be able to:

- Perform a dc analysis on a gate-bias circuit.
- Perform a dc analysis on a self-bias circuit.
- Demonstrate the bias instability of both types of bias circuits.

DISCUSSION

In Exercise 36, you saw that JFET drain current can be controlled by varying the amount of reverse bias applied to the gate-source junction of the device. In this exercise, you will see how two JFET biasing circuits develop the value of V_{GS} needed to establish the JFET operating point.

Gate bias uses a negative voltage source (connected to the gate) to reverse bias the gate-source junction. *Self-bias,* on the other hand, is designed so that the gate is held at ground potential, while a positive voltage is developed across a source resistor (R_S). With the gate grounded and the source at some *positive* potential, a *negative* V_{GS} is produced, and the Q-point is established.

The primary drawback to using either of these biasing circuits is that they both provide extremely unstable Q-points. In this exercise, you will investigate the Q-point instability and dc operating principles of these JFET biasing circuits.

LAB PREPARATION

Review:

- Sections 21.1 and 21.2 of *Electronics Technology Fundamentals*.
- The specification sheet for the 2N5486 JFET.

MATERIALS

1   dual-polarity variable dc power supply
2   DMMs
1   protoboard
2   resistors: 4.7 kΩ and 1 MΩ
2   potentiometers: 2 kΩ and 5 kΩ (precision)
2   2N5486 *n*-channel JFETs

PROCEDURE

1. Refer to the spec sheet for the 2N5486, and determine the ranges of I_{DSS} and V_P for the device. Record these values below.

$$V_P = \underline{\hspace{2cm}} \text{ to } \underline{\hspace{2cm}}$$

$$I_{DSS} = \underline{\hspace{2cm}} \text{ to } \underline{\hspace{2cm}}$$

> *Note:* V_P is the positive equivalent of the $V_{GS(\text{off})}$ rating.

2. Use the values from Step 1 and the appropriate equation to solve for the range of I_D at $V_{GS} = -1$ V, and record these values below.

$$I_D = \underline{\hspace{2cm}} \text{ to } \underline{\hspace{2cm}} \quad @ \ V_{GS} = -1 \text{ V}$$

3. Construct the circuit shown in Figure 37.1a. Initially, R_D should be set to its maximum value and V_{GG} to its minimum.

> *Note:* It is important that you take your current readings in this step as quickly as possible before the circuit heats up.

4. Apply power, and adjust V_{GG} so that $V_{GS} = -1$ V. Then, adjust R_D so that the circuit is midpoint biased ($V_{DS} = \frac{1}{2} V_{DD}$). Use your DMM to measure I_D with V_{GS} set to -1 V, and record both values below.

$$I_D = \underline{\hspace{2cm}} \quad V_{DS} = \underline{\hspace{2cm}}$$

5. Allow the circuit to heat up for a few minutes. Then, measure I_D and V_{DS} again, and record the values below.

$$I_D = \underline{\hspace{2cm}} \quad V_{DS} = \underline{\hspace{2cm}}$$

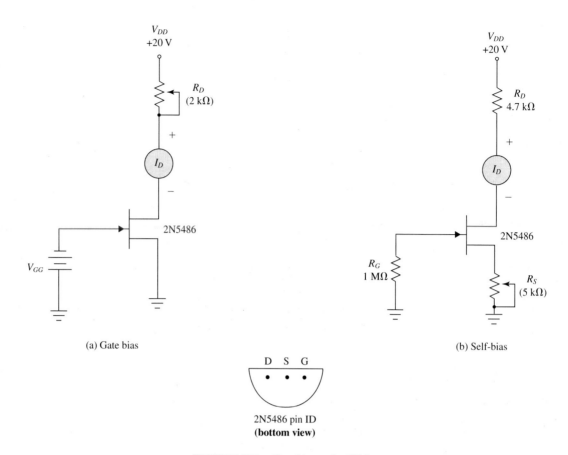

FIGURE 37.1 Gate bias and self-bias.

6. Disconnect power and swap the 2N5486 JFETs. Restore power and adjust R_D, if necessary, so that the circuit is once again midpoint biased. Measure I_D and V_{DS}. Record the values below.

$$I_D = \text{_____} \qquad V_{DS} = \text{_____}$$

7. Using the values from Step 1, plot the maximum and minimum transconductance curves for the 2N5486 in Figure 37.2.

8. Now, plot the dc bias line for the circuit shown in Figure 37.1b on your transconductance curves. Assume $R_S = 1\text{ k}\Omega$ and any value of V_{GS} (other than zero) to plot the bias line. (Example 21.5 of *Electronics Technology Fundamentals* demonstrates the process for plotting the dc bias line.)

9. Determine the ranges of V_{GS} and I_D for the circuit using the Q_{max} and Q_{min} points on your dc bias line. Record these values below.

$$V_{GS} = \text{_____} \text{ to } \text{_____}$$
$$I_D = \text{_____} \text{ to } \text{_____}$$

10. Construct the circuit shown in Figure 37.1b. Adjust R_S so that the circuit is midpoint biased, then measure and record the following values:

$$V_{GS} = \text{_____} \qquad I_D = \text{_____} \qquad V_{DS} = \text{_____}$$

11. Disconnect power, and swap the 2N5486 JFETs. Restore power, and measure and record the following values:

$$V_{GS} = \text{_____} \qquad I_D = \text{_____} \qquad V_{DS} = \text{_____}$$

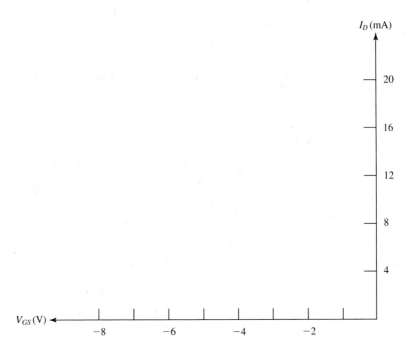

FIGURE 37.2 The 2N5486 transconductance curves.

QUESTIONS & PROBLEMS

1. Refer to Steps 2 and 4. Did your measured value of I_D fall within the range that you calculated in Step 2? If not, explain the discrepancy.

2. Refer to Steps 4 and 5. What changes did you observe as the circuit warmed up? Why do you think these changes occurred?

3. Refer to Steps 10 and 11. What do your results tell you about the bias stability of the self-bias circuit?

4. Refer to Steps 10 and 11. Plot the points that represent the combinations of I_D and V_{GS} from these two steps in Figure 37.2. Do these points fall on the dc bias line? If not, how would you explain that they don't? (*Hint:* What was the assumed value of R_S?)

SIMULATION EXERCISE

Discussion

It has been stated that one of the primary differences between FETs and BJTs is that FETs are *voltage-controlled devices* while BJTs are *current-controlled devices*. In earlier exercises, you saw that BJT collector current (I_C) is dependent on base current (I_B). In the first part of this simulation exercise, you will see that JFET drain current (I_D) is independent of gate current (I_G).

As was the case with BJTs, there are a variety of biasing techniques that can be used with JFETs. When choosing one biasing technique over another, several things must be taken into consideration, such as cost, complexity, etc. One of the most important considerations is stability. You investigated Q-point stability in BJT bias circuits in Exercises 31 and 32. In the second part of this exercise, you will look at the Q-point stability of gate-bias and self-bias circuits.

Procedure

1. Open file Ex37.1 from the Electronics Technology Fundamentals companion web site (www.prenhall.com/paynter).
2. Run the simulation on the gate-bias circuit. Set V_{GS} to 0 V. Measure and record the values of I_D, V_{DS}, and I_G in Table 37.1. Then, increase V_{GS} to −4 V, and repeat the measurements listed at this value of V_{GS}.

TABLE 37.1

V_{GS}	V_{DS} (Measured)	I_D (Measured)	I_G (Measured)
0 V			
−4 V			

3. Run the simulation of the self-bias circuit. First, set the source potentiometer (R_S) to its minimum value (0 Ω), and record the values of I_D, V_{GS} and I_G in Table 37.2. Repeat these measurements for R_S values of 2.5 kΩ and 5 kΩ. Record your results in Table 37.2.

TABLE 37.2

R_S	V_{GS} (Measured)	I_D (Measured)	I_G (Measured)
0 Ω			
2.5 kΩ			
5 kΩ			

4. Adjust R_S so that the self-bias circuit is midpoint biased. Measure and record the values of V_{GS}, V_{DS}, and I_D in Table 37.3.

TABLE 37.3

Device	V_{GS} (Measured)	I_{DS} (Measured)	I_D (Measured)
2N5486			
2N5485			
2N5484			

5. Replace the 2N5486 with the other two devices listed in Table 37.3. These three FETs are from the same family, but have different I_{DSS} and $V_{GS(off)}$ ratings. For each device, repeat the measurements from Step 4 and record your results in Table 37.3.

1. Refer to your results in Steps 2 and 3. Use these results to identify the JFET as a current-controlled or a voltage-controlled device.

2. Refer to your results from Step 3. Based on your knowledge of how the self-bias circuit works, explain your results for the three values of R_S.

3. Refer to your results from Steps 4 and 5. What do these results tell you about the Q-point stability of these two bias circuits?

Exercise 38

The Common-Source Amplifier

OBJECTIVES

After completing this exercise, you should be able to:

- Analyze the ac operation of a typical common-source amplifier.
- Demonstrate the differences between the ac operating characteristics of typical common-source and common-emitter amplifiers.

DISCUSSION

There are several similarities between *common-source (CS)* and *common-emitter (CE)* amplifiers. Both provide a measurable amount of voltage gain, and both have a 180° voltage phase shift between their input and output signals. At the same time, however, they also have several differences.

Perhaps the biggest difference is that JFETs are voltage-controlled devices and BJTs are current-controlled devices. Also, the CS amplifier typically has much higher input impedance than the CE amplifier. Finally, the voltage gain calculation for a CS amplifier is very different from that for a CE amplifier.

In this exercise, you will observe the operation of a self-biased CS amplifier. While analyzing this exercise, you should pay close attention to those points of operation that distinguish the CS from the CE amplifier.

- Review Section 21.3 of *Electronics Technology Fundamentals*.
- Review the specification sheet for the 2N5486 JFET.
- Review the discussion on fault analysis in Appendix D.
- Make six copies of the *fault analysis chart* in Appendix D (for the fault simulations portion of the exercise).

MATERIALS

1	variable dc power supply
1	function generator
1	dual-trace oscilloscope
2	DMMs
1	protoboard
3	resistors: 2.2 kΩ, 4.7 kΩ, and 1 MΩ
2	potentiometers: 5 kΩ and 2 MΩ (precision)
3	capacitors: 1 μF (2) and 22 μF
1	2N5486 *n*-channel JFET

PROCEDURE

1. Construct the circuit shown in Figure 38.1. Place one of your DMMs in the drain circuit to measure I_D. The other meter will be used to make voltage measurements. R_S should be set initially to 1 kΩ.

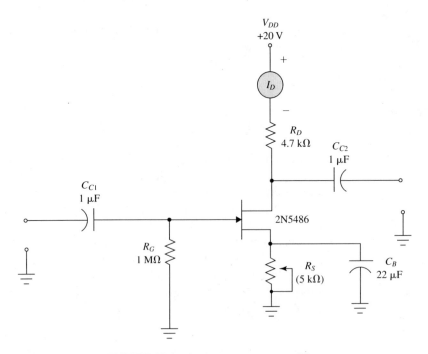

FIGURE 38.1   A common-source amplifier.

2. Measure the following values:

$$V_{GS} = \underline{\hspace{2cm}} \qquad I_D = \underline{\hspace{2cm}}$$

3. Change R_S to 1.5 kΩ, and measure the following values:

$$V_{GS} = \underline{\hspace{2cm}} \qquad I_D = \underline{\hspace{2cm}}$$

4. Use your results from Steps 2 and 3 to calculate ΔV_{GS} and ΔI_D. Then, use these values to calculate g_m. Record your results below.

$$\Delta V_{GS} = V_{GS(max)} - V_{GS(min)} = \underline{\hspace{2cm}}$$

$$\Delta I_D = I_{D(max)} - I_{D(min)} = \underline{\hspace{2cm}}$$

$$g_m = \frac{\Delta I_D}{\Delta V_{GS}} = \underline{\hspace{2cm}}$$

5. Using the rated value of R_D and the value of g_m found in Step 4, calculate the open-load voltage gain of the amplifier.

$$A_v = \underline{\hspace{2cm}}$$

6. Connect the function generator and oscilloscope to the circuit as shown in Figure 38.2. Set the output amplitude of the function generator to minimum and the frequency to 5 kHz. Leave R_S set to 1.5 kΩ. Note that the ammeter has been removed from the circuit.

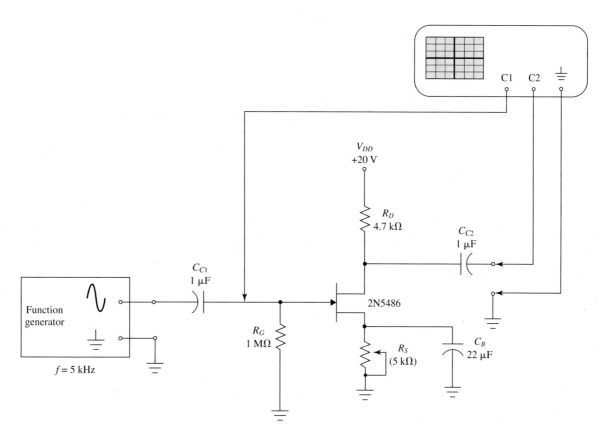

FIGURE 38.2

7. Slowly increase the input amplitude until you get the maximum undistorted output signal from the amplifier. Note the phase relationship between the input and output waveforms in the space provided below.

8. Measure and record the peak-to-peak values of the input and output waveforms. Then, use these values to determine the actual voltage gain of the amplifier.

$$v_{in} = \underline{\hspace{2cm}}$$

$$v_{out} = \underline{\hspace{2cm}}$$

$$A_v = \frac{v_{out}}{v_{in}} = \underline{\hspace{2cm}}$$

9. Connect a 2.2 kΩ load to the amplifier, and repeat Steps 7 and 8.

$$v_{in} = \underline{\hspace{2cm}}$$

$$v_{out} = \underline{\hspace{2cm}}$$

$$A_v = \frac{v_{out}}{v_{in}} = \underline{\hspace{2cm}}$$

> *Note:* In Step 10, you will employ a procedure similar to that used in previous labs to determine the input impedance of an amplifier. The only difference is that you will monitor the *output* of the amplifier. When the output of the amplifier drops to one-half its normal output, this means that one-half the input signal is dropped across the series potentiometer.

10. Connect a 2 MΩ potentiometer between the function generator and the input coupling capacitor. Initially, the pot should be at its lowest setting. Measure the input impedance of the amplifier as follows:

   • Adjust the input so that you get the maximum undistorted output signal.
   • Slowly increase the resistance of the potentiometer until v_{out} is reduced to one-half its original value.
   • Disconnect power from the circuit, and carefully remove the potentiometer. Measure its adjusted resistance, and record this value below.

$$R \cong Z_{in} = \underline{\hspace{2cm}}$$

1. Refer to Steps 5 and 8. How did your measured value of A_v compare to the calculated value? If there is a significant difference, how do you account for this variation?

2. What happened to the voltage gain of the amplifier when the load was connected in Step 9? Explain why this happened.

3. Refer to Step 7. What was the phase relationship between the input and output waveforms? Which BJT amplifier has the same input/output phase relationship?

4. Refer to your results from Step 10. How close is your measured value of Z_{in} to the value of the gate resistor? What effect does the input impedance of the JFET have on the circuit input impedance?

SIMULATION EXERCISE

Procedure

1. Open file Ex38.1 from the Electronics Technology Fundamentals companion web site (www.prenhall.com/paynter).
2. Run the simulation and make certain that the circuit is operating properly. Record the measurements listed below so that you have a reference for an unfaulted circuit.

$$V_{DS} = \underline{\hspace{2cm}}$$

$$V_{GS} = \underline{\hspace{2cm}}$$

$$A_v = \underline{\hspace{2cm}}$$

3. Run each of the fault simulations listed below. Record your results/analysis for each fault in one of the fault analysis charts. Remember to return each component to its no-fault condition before introducing the next fault.

> *Note:* Some faults that are inserted using the simulator can be dangerous to equipment and/or personnel when they occur in real circuits. Simulations provide exposure to such faults in a safe environment.

Fault Simulations

Fault 38-1   Shorted gate resistor (R_G).

Fault 38-2   Open JFET source terminal.

Fault 38-3   Shorted JFET (gate-to-source).

Fault 38-4   Shorted JFET (drain-to-source).

Fault 38-5   Shorted bypass capacitor.

Fault 38-6   Open source resistor (R_S).

Exercise 39

Inverting Amplifiers

OBJECTIVES

After completing this exercise, you should be able to:

- Analyze the operation of the inverting amplifier.
- Determine the closed-loop voltage gain of an inverting amplifier.
- Demonstrate the effects of load resistance on amplifier gain.

DISCUSSION

The inverting amplifier is very similar in several respects to the CE and CS amplifiers. Like those circuits, the inverting amplifier produces a 180° voltage phase shift from its input to its output. Inverting amplifiers can also be designed for a wide range of voltage gains. At the same time, the inverting amplifier has many characteristics that make it more desirable than either the CE or CS amplifier:

- Inverting amplifiers are capable of extremely high voltage gains—well into the thousands.
- The gain of an inverting amplifier is extremely stable and easy to calculate.
- Inverting amplifiers are easier to design and troubleshoot than CE or CS amplifiers.

Since this is probably your first exposure to working with operational amplifiers (op-amps), here are a few points to keep in mind:

- The $+V$ and $-V$ pins *must* be connected to their respective supply voltages for the op-amp to work.
- Pin 1 is located next to the indentation in the integrated circuit (IC) package, as shown in Figure 39.1. The rest of the pins are numbered in sequence, going counterclockwise from pin 1.

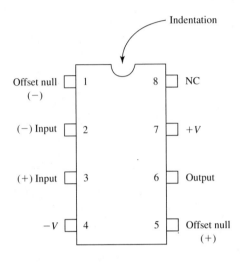

Note: NC means *Not Connected.*

FIGURE 39.1 The pin configuration of the 741 op-amp.

- The op-amp will not work if either the inverting (pin 2) or the noninverting (pin 3) inputs are left open.
- Never apply an input signal to an op-amp unless both supply voltages are connected to the IC.

LAB PREPARATION

- Review Sections 22.1 through 22.4 of *Electronics Technology Fundamentals*.
- Review the specification sheet for the 741 op-amp.
- Review the discussion on fault analysis in Appendix D.
- Make six copies of the *fault analysis chart* in Appendix D (for the fault simulations portion of the exercise).

MATERIALS

1   dual-polarity variable dc power supply
1   function generator
1   dual-trace oscilloscope

1   protoboard
9   resistors: 470 Ω, 1 kΩ, 10 kΩ (3), 27 kΩ, 39 kΩ, 47 kΩ, and 82 kΩ
1   1 kΩ potentiometer
1   KA741 op-amp (or equivalent)

PROCEDURE

1. Calculate the *closed-loop voltage gain* for the circuit shown in Figure 39.2.

$$A_{CL} = \frac{R_f}{R_{in}} = \underline{\hspace{2cm}}$$

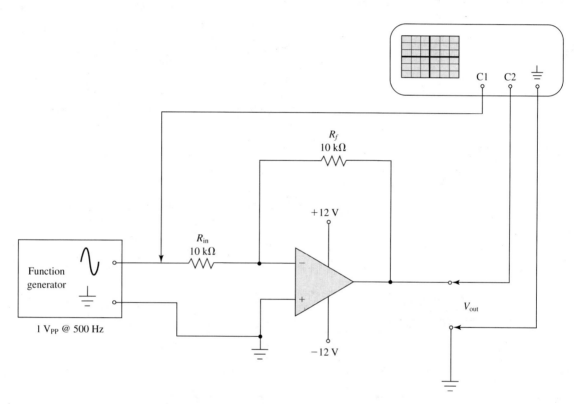

FIGURE 39.2   An inverting amplifier.

2. Construct the circuit shown in Figure 39.2. (*Note:* If you are using a function generator with dc offset controls, make sure that the offset is either set at 0 V or disabled.)
3. Measure the peak-to-peak values of the input and output waveforms, and record their values below. Then, draw both waveforms in Figure 39.3.

$$v_{in} = \underline{\hspace{2cm}} \qquad v_{out} = \underline{\hspace{2cm}}$$

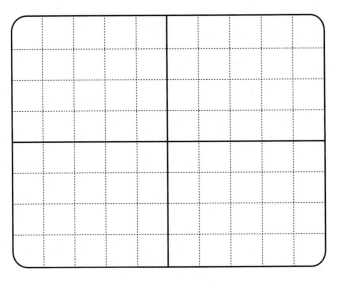

FIGURE 39.3 Inverting amplifier waveforms.

4. Use your results from Step 3 to calculate the closed-loop voltage gain for the circuit.

$$A_{CL} = \frac{v_{\text{out}}}{v_{\text{in}}} = \underline{\hspace{3cm}}$$

5. Table 39.1 lists a series of resistance values to be used in place of R_f. For each value listed, repeat the measurements and calculations in Steps 1 through 4.

TABLE 39.1 **Voltage Measurements and Calculations**

R_f	v_{in}	v_{out}	A_{CL} *(Calculated)*	A_{CL} *(Measured)*
27 kΩ				
39 kΩ				
47 kΩ				
82 kΩ				

6. Return R_f to 10 kΩ. Connect a 10 kΩ resistor as a load, and increase the input signal to 5 V$_{pp}$. Measure the output voltage across the load. Use this value to calculate A_{CL}. Record these values in Table 39.2.
7. Repeat Step 6 for the other two load values listed in Table 39.2.

TABLE 39.2 **Voltage Measurements and Calculations**

R_L	v_{in}	v_{out}	A_{CL}
10 kΩ			
1 kΩ			
470 Ω			

8. Connect the 1 kΩ potentiometer as a load. It should initially be adjusted to its highest setting. Slowly decrease the value of the potentiometer until you begin to see a decrease in output voltage or the output waveform begins to distort. Disconnect power from the circuit, and carefully remove the potentiometer. Measure the resistance of the pot, and record this value below.

$$R_L = _____ \ \Omega$$

QUESTIONS & PROBLEMS

1. Refer to Figure 39.3. What was the phase relationship between the input and the output signals? Explain this phase relationship.

2. Refer to your results from Steps 1 and 4. How did your measured value of A_{CL} compare to the calculated value? How would you account for any significant difference between these two values?

3. Refer to Table 39.2. What effect did a change in load have on the voltage gain of the amplifier? How does this compare to CE and CS amplifiers?

4. Refer to Step 8. Based on your knowledge of op-amp output impedance, explain why the output waveform began to change at this value of R_L.

SIMULATION EXERCISE

Procedure

Open file Ex39.1 from the Electronics Technology Fundamentals companion web site (www.prenhall.com/paynter). This is the same circuit as the one shown in Figure 39.2. Run the simulation. When you are satisfied that the circuit is operating normally, continue with the fault simulation portion of the exercise.

> *Note:* Some faults that are inserted using the simulator can be dangerous to equipment and/or personnel when they occur in real circuits. Simulations provide exposure to such faults in a safe environment.

Fault Simulations

Fault 39-1   Open op-amp input (Pin 2).

Fault 39-2   Open op-amp input (Pin 3).

Fault 39-3   Open feedback resistor (R_f).

Fault 39-4   Shorted feedback resistor (R_f).

Fault 39-5   Shorted input resistor (R_{in}).

Fault 39-6   Open input resistor (R_{in}).

Exercise 40

Noninverting Amplifiers

OBJECTIVES

After completing this exercise, you should be able to:

- Analyze the operation of the noninverting amplifier.
- Determine and measure the closed-loop voltage gain of a noninverting amplifier.
- Demonstrate the relationship between gain and bandwidth for operational amplifiers.

DISCUSSION

The noninverting amplifier shares many characteristics with the inverting amplifier, with two exceptions:

- As the name implies, the output of this amplifier is in phase with its input.
- The noninverting amplifier has significantly higher input impedance than a comparable inverting amplifier.

As shown in Figure 40.1, the input is connected directly to the noninverting terminal of the op-amp. As a result, the circuit input impedance is equal to (or greater than) the input impedance of the op-amp itself.

When compared to discrete amplifier circuits (like the emitter or source follower), the noninverting amplifier shares some of their characteristics as well. It has high input

321

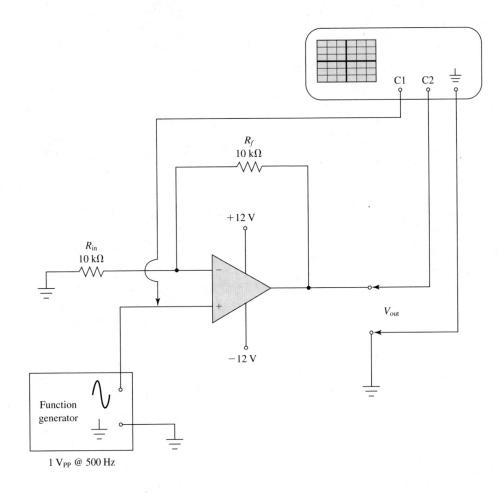

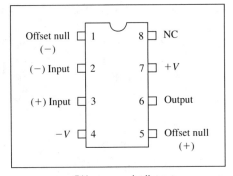

741 op-amp pin diagram

FIGURE 40.1   A noninverting amplifier.

impedance, low output impedance, and the input and output signals are in phase. The one major difference is that noninverting amplifiers can be designed for high voltage gain, whereas emitter and source followers are limited to voltage gains of slightly less than unity.

In this exercise, you will investigate the basic operation of the noninverting amplifier. You will also use the noninverting amplifier to look briefly at the relationship between gain and bandwidth for operational amplifiers.

- Review Sections 22.5 and 22.6 of *Electronics Technology Fundamentals*.
- Review the specification sheet for the 741 op-amp.
- Review the discussion on fault analysis in Appendix D.
- Make six copies of the *fault analysis chart* in Appendix D (for the fault simulations portion of the exercise).

MATERIALS

1   dual-polarity variable dc power supply
1   function generator
1   dual-trace oscilloscope
1   protoboard
6   resistors: 10 kΩ (2), 27 kΩ, 39 kΩ, 47 kΩ, and 82 kΩ
1   KA741 operational amplifier (or equivalent)

PROCEDURE

1. Calculate the closed-loop voltage gain for the circuit shown in Figure 40.1.

$$A_{CL} = \frac{R_f}{R_{in}} + 1 = \underline{\hspace{2cm}}$$

2. Construct the circuit shown in Figure 40.1. (*Note:* If you are using a function generator with dc offset controls, make sure that the offset is at 0 V or disabled.)
3. Measure the peak-to-peak values of the input and output waveforms, and record these values below. Draw both waveforms in Figure 40.2.

$$v_{in} = \underline{\hspace{2cm}} \qquad v_{out} = \underline{\hspace{2cm}}$$

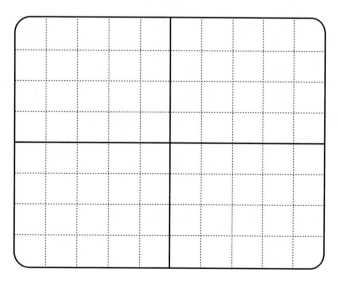

FIGURE 40.2   Noninverting amplifier input and output waveforms.

4. Use your results from Step 3 to calculate the closed-loop voltage gain for this circuit.

$$A_{CL} = \frac{v_{out}}{v_{in}} = \underline{\hspace{2cm}}$$

5. Table 40.1 lists a series of resistance values to be used in place of R_f. For each value listed, repeat the measurements and calculations in Steps 1 through 4.

TABLE 40.1 Voltage Measurements and Calculations

R_f	v_{in}	v_{out}	A_{CL} (Calculated)	A_{CL} (Measured)
27 kΩ				
39 kΩ				
47 kΩ				
82 kΩ				

6. Return R_f to 27 kΩ. Now, slowly increase the input frequency until the output signal starts to become distorted. Record this frequency below, and draw the input and output waveforms in Figure 40.3a.

$$f_{max} = \underline{\hspace{2cm}} @ R_L = 27 \text{ k}\Omega$$

7. Change R_f to 47 kΩ, and repeat Step 6. Draw the waveforms in Figure 40.3b.

$$f_{max} = \underline{\hspace{2cm}} @ R_L = 47 \text{ k}\Omega$$

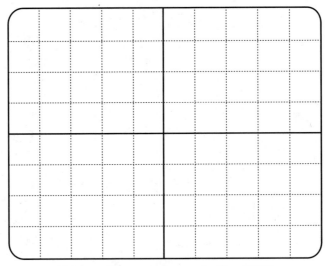

(a) Input and output waveforms at f_{max} (R_L = 27 kΩ)

FIGURE 40.3 Waveforms for Steps 6 and 7.

(continues)

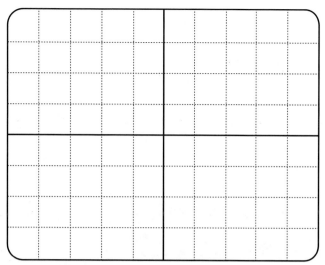

(b) Input and output waveforms at f_{max} (R_L = 47 kΩ)

FIGURE 40.3   *(continued)*

QUESTIONS & PROBLEMS

1. Refer to Steps 1 through 5. How did your measured values of A_{CL} compare with the calculated values? Explain any difference between the measured and calculated values.

2. Refer to Step 6. Calculate the maximum operating frequency using the op-amp's slew rate (from the spec sheet). How does this value compare with your measured value? Explain any difference between your calculated and measured values.

3. Refer to Steps 6 and 7. Did the maximum frequency increase or decrease when R_f was increased? Explain why f_{max} did (or did not) change.

SIMULATION EXERCISE

Procedure

Open file Ex40.1 from the Electronics Technology Fundamentals companion web site (www.prenhall.com/paynter). This is the same circuit as the one shown in Figure 40.1. Run the simulation. When you are satisfied that the circuit is operating normally, continue with the fault simulation portion of the exercise.

> *Note:* Some faults that are inserted using the simulator can be dangerous to equipment and/or personnel when they occur in real circuits. Simulations provide exposure to such faults in a safe environment.

Fault Simulations

Fault 40-1   Open op-amp input (Pin 2).

Fault 40-2   Open op-amp input (Pin 3).

Fault 40-3   Open feedback resistor (R_f).

Fault 40-4   Shorted feedback resistor (R_f).

Fault 40-5   Shorted input resistor (R_{in}).

Fault 40-6   Open input resistor (R_{in}).

Exercise 41

Comparators and Summing Amplifiers

OBJECTIVES

After completing this exercise, you should be able to:

- Analyze the operation of a comparator.
- Analyze the operation of a summing amplifier.

DISCUSSION

Although both of the circuits in this exercise are built around op-amps, there is no other relationship between them. You will study them together here simply because it is convenient to do so.

A *comparator* is a circuit used to compare two input voltages and to provide a dc output that indicates which of the two inputs is greater. In most cases, the comparator is used to compare a changing input voltage to a set dc reference voltage. The dc outputs from a comparator are usually referred to as *high* and *low*. In the first part of this exercise, you will look at the basic operation of an op-amp comparator.

The *summing amplifier* is a circuit that produces an output that is proportional to the sum of its inputs. The term *proportional* is used because the circuit may—or may not—have gain, and it produces an inverted output. The inputs can be dc or ac values. In the second part of this exercise, you will examine the operation of a two-input summing amplifier using sine-wave inputs.

LAB PREPARATION

- Review Section 22.7 of *Electronics Technology Fundamentals*.
- Review the specification sheet for the 741 op-amp.
- Review the discussion on fault analysis in Appendix D.
- Make five copies of the *fault analysis chart* in Appendix D (for the fault simulations portion of the exercise).

MATERIALS

1	dual-polarity variable dc power supply
1	function generator
1	dual-trace oscilloscope
1	protoboard
5	resistors: 1 kΩ, 10 kΩ (3), and 22 kΩ
2	KA741 operational amplifiers (or equivalent)

PROCEDURE

Part 1: Comparators

1. Construct the circuit shown in Figure 41.1a. Adjust the function generator to produce a 5-V_{PP} sine wave at a frequency of 1 kHz.
2. Set up your oscilloscope so that both traces are ground-referenced to the center of the display. (This will enable you to use your oscilloscope to measure the value of input voltage that causes the op-amp output to change states.) To establish the center of the grid as the ground reference:

   • Set both AC/GND/DC switches to the ground (GND) position.

   • Move both traces to the center of the display.

   • Set both switches to the DC position to provide dc coupling.
3. Observe the circuit input and output with the same vertical sensitivity (V/Div). Determine the magnitude of the input voltage at which the output changes state, and record this value below.

   V_{in} = _____ when V_{out} changes state
4. Draw both waveforms in Figure 41.2a.
5. Modify the circuit as shown in Figure 41.1b. Repeat Steps 3 and 4. Record the reference voltage value below, and draw the waveforms in Figure 41.2b.

   V_{in} = _____ when V_{out} changes state

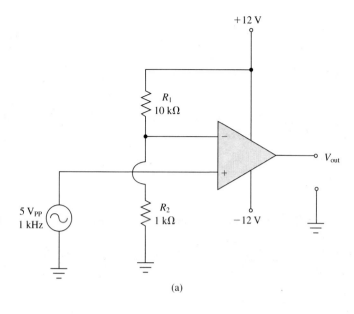

(a)

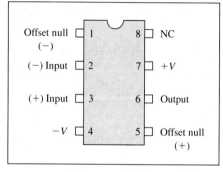

741 op-amp pin diagram

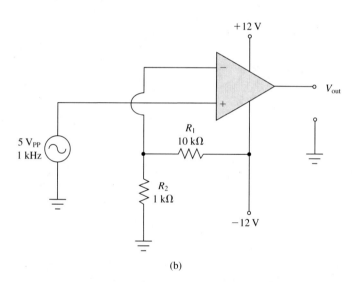

(b)

FIGURE 41.1 Comparators.

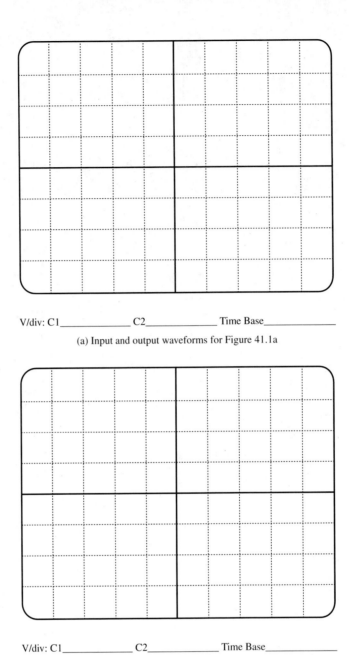

V/div: C1_____ C2_____ Time Base_____

(a) Input and output waveforms for Figure 41.1a

V/div: C1_____ C2_____ Time Base_____

(b) Input and output waveforms for Figure 41.1b

FIGURE 41.2 Comparator waveforms.

Part 2: Summing Amplifiers

6. Construct the circuit shown in Figure 41.3. Note that the signal generator is supplying both inputs to the summing amplifier, which is op-amp 2. Op-amp 1 in Figure 41.3 is being used to provide isolation between the two summing amplifier inputs.

7. Measure and record the peak-to-peak output voltage.

$$V_{\text{out}} = \underline{\hspace{2cm}}$$

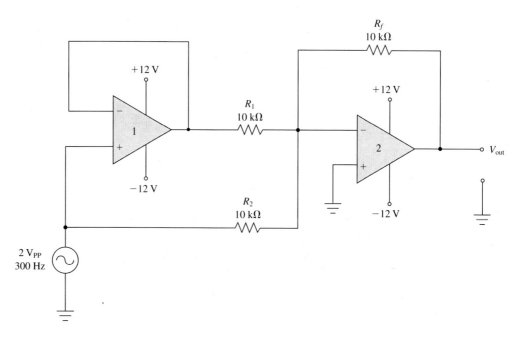

FIGURE 41.3   A two-input summing amplifier.

8. Change R_1 to 22 kΩ, and measure the peak-to-peak output voltage.

$$V_{out} = \underline{\hspace{2cm}}$$

9. Restore R_1 to 10 kΩ, and change R_f to 22 kΩ. Measure the peak-to-peak output voltage.

$$V_{out} = \underline{\hspace{2cm}}$$

QUESTIONS & PROBLEMS

1. Using the nominal values of the resistors shown in Figure 41.1a, calculate V_{ref}. How does this value compare to the voltage measured in Step 3? Explain any discrepancy between the two values.

$$V_{ref} = +V\frac{R_2}{R_1 + R_2} = \underline{\hspace{2cm}}$$

2. Refer to the waveforms you recorded in Figure 41.2. Explain the input/output relationship that you observed for these two circuits.

3. Refer to Step 8. Identify and explain any effect that increasing R_1 had on the circuit's input/output relationship.

4. Refer to Step 9. Identify and explain any effect that increasing R_f had on the circuit's input/output relationship.

SIMULATION EXERCISE

Procedure

Open files Ex41.1 and Ex41.2 from the Electronics Technology Fundamentals companion web site (www.prenhall.com/paynter). File Ex41.1 contains the two comparator circuits from Figure 41.1. File Ex41.2 contains the summing amplifier circuit from Figure 41.3. When you open the files, note the tabs on the lower left-hand portion of your screen. By clicking on these tabs, you can view either of the two files that you have opened.

Begin with the two comparator circuits. Run the simulation. When you are satisfied that the circuits are operating normally, continue with the fault simulation portion of the exercise.

When you have completed the first three fault analysis charts, change to File Ex41.2. Run the simulation. Make certain that the circuit is working properly then complete the last two fault simulations on the summing amplifier circuit.

> *Note:* Some faults that are inserted using the simulator can be dangerous to equipment and/or personnel when they occur in real circuits. Simulations provide exposure to such faults in a safe environment.

Fault Simulations

Fault 41-1   Open R_1 (Circuit 41.1a).

Fault 41-2   Open R_2 (Circuit 41.1a).

Fault 41-3   Shorted R_1 (Circuit 41.1b).

Fault 41-4   Open R_f (Circuit 41.2).

Fault 41-5   Open R_1 (Circuit 41.2).

Exercise 42

Low-Pass and High-Pass Active Filters

OBJECTIVES

After completing this exercise, you should be able to:

- Analyze the basic operation of a single-pole and two-pole low-pass filter.
- Analyze the basic operation of a single-pole and two-pole high-pass filter.

DISCUSSION

Active filters are tuned op-amp circuits that contain one or more poles. A *pole* is simply a single *RC* circuit.

The roll-off rate for an active filter depends on the number of poles it contains. For example, a Butterworth active filter has a roll-off rate of 6 dB per octave (20 dB per decade) per pole. Thus, a two-pole Butterworth filter has a roll-off rate of approximately 12 dB per octave (40 dB per decade).

Low-pass active filters are designed to pass all frequencies *below* a predetermined cutoff frequency (f_C). In the first part of this exercise, you will observe the operation of a single-pole low-pass filter and a two-pole low-pass filter.

High-pass active filters are designed to pass all frequencies *above* a predetermined cutoff frequency (f_C). In the second part of this exercise, you will look at both a single-pole and a two-pole high-pass filter. Note that the same equations are used to determine f_C for both high-pass and low-pass filters.

- Review Sections 23.1 through 23.4 of *Electronics Technology Fundamentals*.
- Review the specification sheet for the 741 op-amp.
- Review the discussion on fault analysis in Appendix D.
- Make eight copies of the *fault analysis chart* in Appendix D (for the fault simulations portion of the exercise).

MATERIALS

1	dual-polarity variable dc power supply
1	function generator
1	dual-trace oscilloscope
1	protoboard
5	resistors: 10 kΩ, 15 kΩ (2), 22 kΩ, and 30 kΩ
3	capacitors: 0.01 µF (2) and 0.022 µF
1	KA741 operational amplifiers (or equivalent)

PROCEDURE

Part 1: Low-Pass Active Filters

1. Calculate the cutoff frequency for the single-pole low-pass filter shown in Figure 42.1.

$$f_C = \frac{1}{2\pi RC} = \underline{\hspace{3cm}}$$

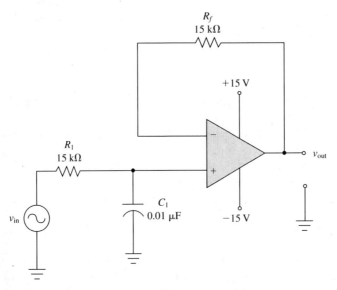

FIGURE 42.1 A single-pole low-pass filter.

2. Construct the circuit shown in Figure 42.1. Connect Channel 1 of the oscilloscope to the input and Channel 2 to the output. Adjust the function generator to produce a 1 V_{PP} sine wave at one-fourth the value of f_C calculated in Step 1.
3. Measure and record the peak-to-peak input and output values in Table 42.1. Then, use these values to calculate A_{CL} and $A_{CL(dB)}$. Record these values in the table.

TABLE 42.1   Voltage and Gain Values for the Circuit in Figure 42.1

f_{in}	v_{in}	v_{out}	A_{CL}	$A_{CL(dB)}$
$\frac{1}{4}f_C$				
f_C				
$2f_C$				
$4f_C$				

4. Set the operating frequency to the value of f_C calculated in Step 2, and repeat the voltage measurements and gain calculations in Step 3. Enter your results in Table 42.1.
5. Repeat Step 4 for the remaining frequencies listed in Table 42.1.
6. Construct the two-pole filter shown in Figure 42.2. Calculate the cutoff frequency for this circuit, and record this value below.

$$f_C = \frac{1}{2\pi\sqrt{R_1R_2C_1C_2}} = \underline{\qquad}$$

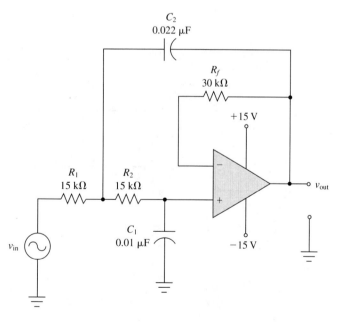

FIGURE 42.2   A two-pole low-pass filter.

7. Repeat Steps 2 through 5 for this circuit, and record your results in Table 42.2.

TABLE 42.2   Voltage and Gain Values for the Circuit in Figure 42.2

f_{in}	v_{in}	v_{out}	A_{CL}	$A_{CL(dB)}$
$\frac{1}{4}f_C$				
f_C				
$2f_C$				
$4f_C$				

Part 2: High-Pass Filters

> *Note:* The second part of this exercise is basically the same as the first. The only difference is that you will analyze a single-stage and a two-stage *high-pass* filter.

8. Calculate the cutoff frequency of the single-stage filter shown in Figure 42.3a.

$$f_C = \underline{\hspace{2cm}}$$

9. Set the input signal to 1 V_{PP} at four times the value of f_C that you calculated in Step 8.

10. Measure the input and output voltages, and use these values to calculate A_{CL} and $A_{CL(dB)}$. Record all these values in Table 42.3.

TABLE 42.3   Voltages and Gain Values for the Circuit in Figure 42.3a

f_{in}	v_{in}	v_{out}	A_{CL}	$A_{CL(dB)}$
$4f_C$				
f_C				
$\frac{1}{2}f_C$				
$\frac{1}{4}f_C$				

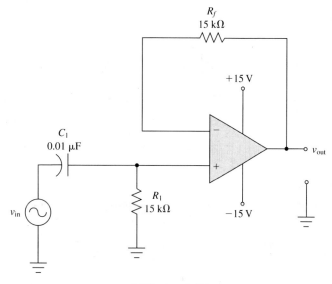

(a) Single-pole filter

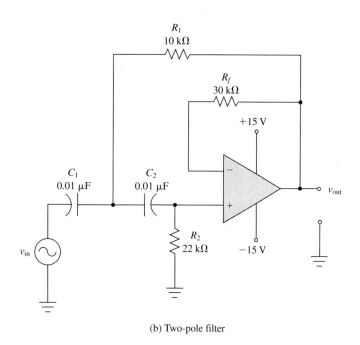

(b) Two-pole filter

FIGURE 42.3   One-pole and two-pole high-pass filters.

11.   Set the operating frequency to the value of f_C calculated in Step 8, and repeat the voltage measurements and gain calculations in Step 10. Enter your results in Table 42.3.

12.   Repeat Step 11 for the remaining frequencies listed in Table 42.3.

13.   Repeat Steps 8 through 12 for the two-stage high-pass filter shown in Figure 42.3b. Record your calculated value for f_C below, and complete Table 42.4.

$$f_C = \underline{\hspace{2cm}}$$

TABLE 42.4   Voltages and Gain Values for the Circuit in Figure 42.3b

f_{in}	v_{in}	v_{out}	A_{CL}	$A_{CL(dB)}$
$4f_C$				
f_C				
$\frac{1}{2}f_C$				
$\frac{1}{4}f_C$				

QUESTIONS & PROBLEMS

1. Compare your measured and calculated cutoff frequencies for all four circuits.
   a. Were there any significant variations between the measured and calculated values for each circuit? If so, identify the circuit and the discrepancy.

   b. Were the variations greater for the single-stage or two-stage filters? Why do you think this was the case?

2. Refer to your results from Tables 42.1 and 42.2. Calculate the roll-off rate for the single-pole and two-pole low-pass filters. Are these roll-off rates consistent with the theoretical values of 6 dB and 12 dB per octave, respectively? If not, explain any discrepancies.

3. Refer to your results from Tables 42.3 and 42.4. Calculate the roll-off rate for the single-pole and two-pole high-pass filters. Are these roll-off rates consistent with the theoretical values of 6 dB and 12 dB per octave, respectively? If not, explain any discrepancies.

SIMULATION EXERCISE

Procedure

1. Open files Ex42.1 and Ex42.2 from the Electronics Technology Fundamentals companion web site (www.prenhall.com/paynter). File Ex42.1 contains the two low-pass filters from Figures 42.1 and 42.2. File Ex42.2 contains the two high-pass filters from Figure 42.3. When you open the files, note the tabs on the lower left-hand portion of your screen. By clicking on these tabs, you can view either of the two files that you have opened.
2. Run the simulation for both circuits. Use the Bode plotters to verify the gain (A_{CL}) and dB gain ($A_{CL(dB)}$) of all the filters. Compare these values to those in Tables 42.1 through 42.4.
3. When you satisfied that the circuits are operating normally, proceed with the fault analysis portion of this exercise.

> *Note:* Some faults that are inserted using the simulator can be dangerous to equipment and/or personnel when they occur in real circuits. Simulations provide exposure to such faults in a safe environment.

Fault Simulations

Fault 42-1 R_1 out of tolerance[a] (Circuit 42.1).

Fault 42-2   Open filter capacitor C_1 (Circuit 42.1).

Fault 42-3   Shorted filter resistor R_2 (Circuit 42.2).

Fault 42-4   Open filter capacitor C_2 (Circuit 42.2).

Fault 42-5   Shorted filter capacitor C_1 (Circuit 42.3a).

Fault 42-6   Open filter resistor R_1 (Circuit 42.3a).

Fault 42-7   Open filter capacitor C_1 (Circuit 42.3b).

Fault 42-8   Shorted filter capacitor C_2 (Circuit 42.3b).

[a]Set leakage to 1.5 kΩ

Exercise 43

Oscillators

OBJECTIVES

After completing this exercise, you should be able to:

- Analyze the operation of the Wien-bridge oscillator.
- Analyze the operation of the discrete Colpitts oscillator.

DISCUSSION

Oscillators can be built around an op-amp or a discrete transistor. In this exercise, you will look at one example of each: the Wien-bridge oscillator is built around an op-amp; the Colpitts oscillator is built around a BJT.

As shown in Figure 43.1, the Wien-bridge oscillator employs both positive and negative feedback. The *positive* feedback network contains two *RC* circuits (one series and one parallel) that determine the circuit operating frequency. The *negative* feedback circuit is used to control the gain of the op-amp.

The Colpitts oscillator is an *LC* oscillator capable of operating at much higher frequencies than a typical *RC* oscillator, such as the Wien-bridge oscillator. The output frequency of the circuit is determined by L, C_1, and C_2, as will be shown in the second part of this exercise.

LAB PREPARATION

Review Sections 23.7 and 23.8 of *Electronics Technology Fundamentals*.

343

1   dual-polarity variable dc power supply

1   function generator

1   dual-trace oscilloscope with a ×10 probe

1   protoboard

7   resistors: 100 Ω, 4.7 kΩ, 15 kΩ (4), and 27 kΩ

2   RF inductors: 1 mH and 10 mH

5   capacitors: 51 pF, 100 pF, 0.01 μF (2), and 0.022 μF

1   50 kΩ potentiometer

1   2N3904 *npn* transistor

1   KA741 operational amplifier (or equivalent)

2   1N4148 small-signal diodes

PROCEDURE

Part 1: The Wien-Bridge Oscillator

1. Construct the circuit shown in Figure 43.1. Set R_4 to 25 kΩ, and connect your oscilloscope to monitor the output of the oscillator.

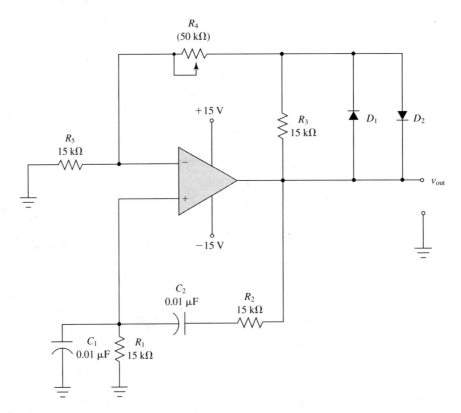

FIGURE 43.1   The Wien-bridge oscillator.

Note: Be sure to use a ×10 oscilloscope probe. This will limit the loading effect of the oscilloscope on circuit operation.

2. Measure the oscillator output frequency, and record this value below.

$$f_{\text{out}} = \underline{\hspace{2cm}}$$

3. Vary the setting of R_4, and note the effect this has on the output waveform. Record your observations below.

4. Remove the diodes from the circuit, and power up. Note the effect on the circuit output, and record your observations below.

5. Vary the setting of R_4 with the diodes removed. Note the effect on the circuit output, and record your observations below.

Part 2: The Colpitts Oscillator

6. Calculate the values below for the oscillator shown in Figure 43.2.

$$C_T = \frac{C_1 C_2}{C_1 + C_2} = \underline{\hspace{2cm}}$$

$$f_{\text{out}} = \frac{1}{2\pi \sqrt{LC_T}} = \underline{\hspace{2cm}}$$

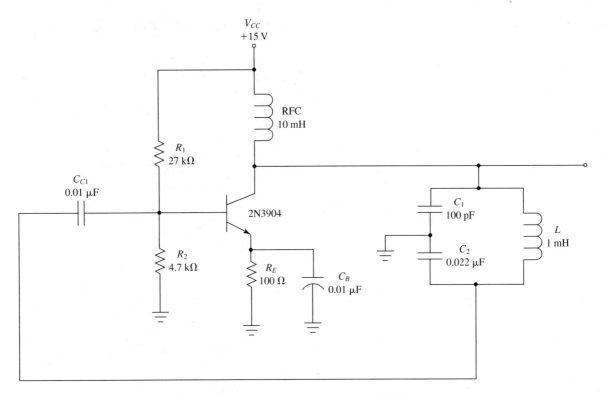

FIGURE 43.2 A discrete Colpitts oscillator.

7. Construct the circuit shown in Figure 43.2, and use the oscilloscope to display the oscillator output signal. Again, it is important that you use the ×10 probe to connect the oscilloscope to the circuit.
8. Measure the oscillator output frequency, and record this value below.

$$f_{out} = \underline{\hspace{2cm}}$$

9. Replace C_1 with a 51 pF capacitor, and repeat Step 8.

$$f_{out} = \underline{\hspace{2cm}}$$

QUESTIONS & PROBLEMS

1. The output frequency of a Wien-bridge oscillator can be approximated using the same equation that you used to calculate the cutoff frequency of a single-pole active filter. By formula,

$$f_{out} = \frac{1}{2\pi RC}$$

Calculate the value of f_{out}, and record this value below.

$$f_{out} = \underline{\hspace{2cm}}$$

How does this value compare to the value measured in Step 2? What do you think could cause any variation between your calculated and measured values of f_{out}?

2. What is the role of the negative feedback path in the Wien-bridge oscillator? Support your answer using your findings in Step 3.

3. Based on your results from Steps 4 and 5, what purpose is served by the diodes in the Wien-bridge oscillator?

4. Compare your calculated and measured values of f_{out} in Steps 6 and 8. Is there any discrepancy between the two? If so, how would you account for this discrepancy?

5. Refer to Steps 8 and 9. Is the output frequency of the Colpitts oscillator related directly or inversely to the capacitance in the resonant circuit? Explain your answer.

SIMULATION EXERCISE

Procedure

1. Open file Ex43.1 from the Electronics Technology Fundamentals companion web site (www.prenhall.com/paynter). This file contains the Wien-bridge oscillator from Figure 43.1. Run the simulation and determine the oscillating frequency of the circuit. Record this value below.

$$f_{out} = \underline{\hspace{2cm}}$$

2. Now, start to decrease the value of the feedback potentiometer (R_4). Note the effect that this has on the oscillator output. As the potentiometer nears 30%, make sure that you wait for the simulator to make several traces of the output waveform. Continue to decrease R_4. Something will occur that takes some time to resolve itself. Record your observations below.

3. Return R_4 to 25 kΩ. Now, change R_1 and R_2 to 10 kΩ. Run the simulation, and determine the output frequency. Record this value below.

$$f_{out} = \underline{\hspace{2cm}}$$

4. Return R_1 and R_2 to 15 kΩ. Now, change C_1 and C_2 to 0.022 μF. Run the simulation, and measure the output frequency. Record this value below.

$$f_{out} = \underline{\hspace{2cm}}$$

Questions

1. Based on your knowledge of the importance of the attenuation factor (α_v) in oscillators, explain your observations in Step 2. What is this type of action called?

2. Based on your results from Step 3, what is the relationship between the resistance in the positive feedback path and the oscillation frequency?

3. Based on your results from Step 4, what is the relationship between the capacitance in the positive feedback path and the oscillation frequency?

Exercise 44

BJT Switching Circuits

After completing this exercise, you should be able to:

- Measure the propagation delay of a BJT switching circuit.
- Identify the contributing factors to propagation delay.
- Demonstrate the effect of a speed-up capacitor on BJT switching time.

Like oscillators, switching circuits can be built around integrated circuits (ICs) or discrete components. In this exercise, you will look at a very simple, BJT-based switching circuit. In Exercise 45, you will look at an op-amp–based switching circuit.

The maximum switching rate of a BJT switch is limited by *delay time (t_d), rise time (t_r), storage time (t_s), and fall time (t_f)*. Delay time and storage time can be drastically reduced by using a *speed-up capacitor*. This capacitor, placed in parallel with the circuit base resistor, provides a high initial reverse bias (reducing storage time) and a high initial forward base current (reducing delay time). In this exercise, you will measure t_d, t_r, t_s, and t_f for a basic BJT switching circuit. You will then add a speed-up capacitor to the circuit and evaluate its effect on the component switching times.

351

Review:

· Sections 24.1 and 24.2 of *Electronics Technology Fundamentals*.
· The specification sheet for the 2N3904 *npn* transistor.

MATERIALS

1   variable dc power supply
1   function generator
1   dual-trace oscilloscope
1   protoboard
2   resistors: 2.2 kΩ and 8.2 kΩ
1   100 pF capacitor
1   2N3904 *npn* transistor

PROCEDURE

1. Construct the circuit shown in Figure 44.1. Connect the oscilloscope so that Channel 1 displays the input waveform and Channel 2 displays the output waveform. Both channels should be dc coupled.

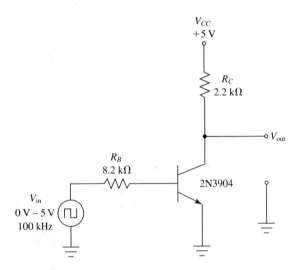

FIGURE 44.1   A BJT switching circuit.

2. Set the function generator for a 0 V to +5 V square-wave output at a frequency of 100 kHz. Draw the input and output waveforms in Figure 44.2.
3. Adjust your oscilloscope controls to produce a display similar to that shown in Figure 44.3, and then measure and record the following values:

$t_d =$ _____ $t_r =$ _____

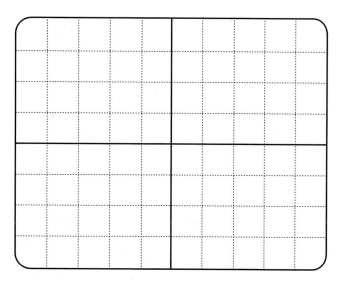

V/div: C1_____ C2_____ Time Base_____

FIGURE 44.2   Switching circuit waveforms.

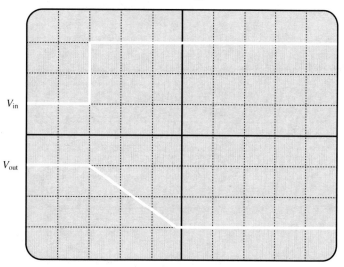

FIGURE 44.3

4. Change your oscilloscope trigger control so that it triggers on the falling edge of the input signal, and then measure and record the following values:

$$t_s = \text{_____} \qquad t_f = \text{_____}$$

5. Disconnect power from the circuit, and install the 100 pF speed-up capacitor in parallel with R_B. Restore power, and repeat Steps 3 and 4. Record your measurements below.

$$t_d = \text{_____} \qquad t_r = \text{_____}$$

$$t_s = \text{_____} \qquad t_f = \text{_____}$$

6. The spec sheet for the 2N3904 lists a rise time rating of 35 ns. From this value, you can calculate the *theoretical* practical upper frequency limit for the 2N3904 as follows:

$$f_{max} = \frac{0.35}{100 t_r} = \frac{0.35}{(100)(35 \text{ ns})} = 100 \text{ kHz}$$

Obviously, you have been operating this device near its limit. Disconnect power from the circuit, remove the speed-up capacitor, and restore power. Now, slowly increase the input frequency above 100 kHz until the leading and trailing edges of the output waveform become rounded. Draw the waveform in Figure 44.4a, and record the frequency at which the waveform rounding was produced. (*Note:* This is a purely subjective measurement. There is no specific correct answer.)

$$f_{max} = \underline{\hspace{2cm}}$$

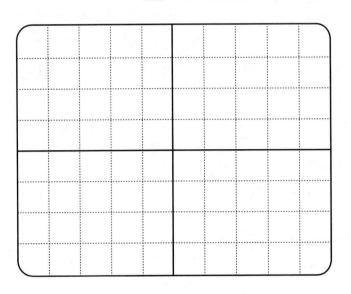

V/div: C1_____ C2_____ Time Base_____

(a) Input and output waveforms at $f > f_{max}$ (speed-up capacitor removed)

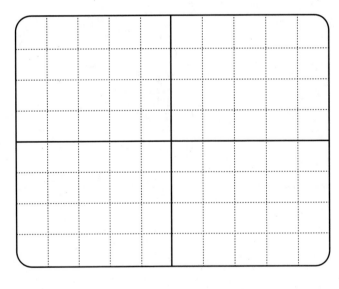

V/div: C1_____ C2_____ Time Base_____

(b) High-frequency distortion waveforms

FIGURE 44.4

7. Reinstall the speed-up capacitor, and repeat Step 6. Draw the waveforms in Figure 44.4b, and record the frequency at which the distorted waveform was produced.

$$f_{max} = \underline{\hspace{2cm}}$$

1. Refer to your results from Steps 3 and 4. Of the four sources of propagation delay, which of these times was the longest? Is this consistent with the statements made in Section 24.2 of the text?

2. Calculate the percent of change in each of your time measurements as a result of installing the speed-up capacitor using:

$$\% \text{ change} = \frac{\text{initial time} - \text{new time}}{\text{initial time}} \times 100$$

% of change in t_d = _____

% of change in t_r = _____

% of change in t_s = _____

% of change in t_f = _____

Are these results consistent with the statements made regarding the results of using a speed-up capacitor in Section 24.2 of the text?

3. Refer to the input and output waveforms that you recorded in Figure 44.2. Why are you viewing the rising edge of the input signal and the falling edge of the output signal at the same time?

4. Refer to Steps 6 and 7. In your opinion, was the actual maximum upper frequency limit close to the one that you calculated? What effect did the speed-up capacitor have on the maximum upper frequency?

SIMULATION EXERCISE

The switching time of various BJT transistors can vary significantly. In this simulation exercise, you will replace the 2N3904 with two other BJT transistors and evaluate their switching characteristics.

Procedure

1. Open file Ex44.1 from the Electronics Technology Fundamentals companion web site (www.prenhall.com/paynter). The file contains two circuits identical to the one in Figure 44.1, except the 2N3904 has been replaced with two different transistors. Begin with the circuit containing the 2SC2001. Run the simulation and measure the delay values listed below. Remember to make your measurements from the 10% and 90% points on the output waveform.

$$t_d = \text{_____} \qquad t_r = \text{_____}$$

$$t_s = \text{_____} \qquad t_f = \text{_____}$$

2. Use your results to calculate the practical upper frequency limit for this switching circuit.

$$f_{max} = \frac{0.35}{100t_r} = \text{_____}$$

3. Repeat Step 1 for the BFS17 transistor.

$$t_d = \text{_____} \qquad t_r = \text{_____}$$

$$t_s = \text{_____} \qquad t_f = \text{_____}$$

4. Repeat Step 2 for this circuit.

$$f_{max} = \frac{0.35}{100t_r} = \text{_____}$$

1. Based on your measurements (in the simulation and in the hardware portions of this exercise), compare the switching capability of the 2SC2001 with that of the 2N3904.

2. Based on your measurements (in the simulation and in the hardware portion of this exercise), compare the switching capability of the BFS17 with that of the 2N3904.

Exercise 45

Op-Amp Schmitt Triggers

OBJECTIVES

After completing this exercise, you should be able to:

- Analyze the basic operation of a Schmitt trigger.
- Demonstrate the concept of symmetrical versus asymmetrical trigger points.
- Determine the upper trigger point (UTP) and lower trigger point (LTP) for symmetrical or asymmetrical Schmitt triggers.

DISCUSSION

As you may remember from Exercise 41, the *comparator* is a voltage-level detector. The comparator is a simple and effective circuit, but it is restricted to *one* set trigger point (reference voltage) and is susceptible to input noise.

A Schmitt trigger is another voltage-level detector. It provides a dc output that changes state when:

- The input makes a *positive-going* transition past a given voltage, called the *upper trigger point (UTP)*.
- The input makes a *negative-going* transition past a given voltage, called the *lower trigger point (LTP)*.

The UTP and LTP voltage values may—or may not—be equal in magnitude, depending on the circuit configuration.

The difference between the UTP and LTP values is called the *hysteresis* value for the circuit. Hysteresis provides built-in noise immunity equal to the difference between the trigger point voltages. This makes the Schmitt trigger more practical than the comparator in noisy environments.

The Schmitt trigger is a *wave-shaping circuit*. No matter what the input looks like, the output will always be in the form of a rectangular wave. The *inverting* Schmitt trigger provides a *low* output (approximately equal to $-V + 1$ V) when the input passes the UTP and a *high* output (approximately equal to $+V - 1$ V) when the input passes the LTP. In this exercise, you will observe the operation of symmetrical and asymmetrical inverting Schmitt triggers.

LAB PREPARATION

Review:

- Section 24.3 of *Electronics Technology Fundamentals*.
- The discussion on fault analysis in Appendix D.
- Make seven copies of the *fault analysis chart* in Appendix D (for the fault simulations portion of the exercise).

MATERIALS

1	dual-polarity variable dc power supply
1	function generator
1	dual-trace oscilloscope
1	DMM
1	protoboard
5	resistors: 1.1 kΩ, 2.2 kΩ, 3.3 kΩ, 10 kΩ, and 100 kΩ
2	1N4148 small-signal diodes
1	KA741 op-amp (or equivalent)

PROCEDURE

Part 1: Symmetrical Trigger Points

1. Calculate the UTP and LTP values for the circuit shown in Figure 45.1, and record these values below.

   UTP = _____   LTP = _____

2. Construct the circuit shown in Figure 45.1. Connect your oscilloscope so that Channel 1 displays the input signal and Channel 2 displays the output signal. Both channels should be dc coupled. Establish the center of the grid as the ground position for both traces.

3. Set the function generator for a 5 V_{PP} sine-wave output at a frequency of 1 kHz. Observe the overlaid input and output waveforms, and draw these waveforms in Figure 45.2.

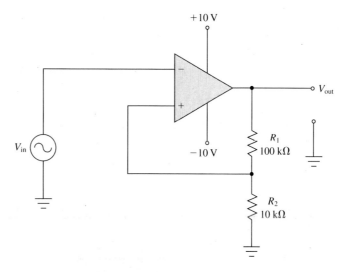

FIGURE 45.1   A symmetrical inverting Schmitt trigger.

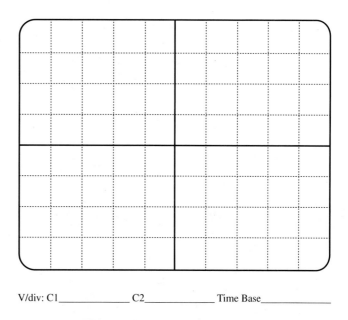

V/div: C1_____ C2_____ Time Base_____

FIGURE 45.2   Waveforms for Step 3.

4.   Use the input and output waveforms to measure the UTP and LTP values for this circuit, as illustrated in Figure 45.3. Record these values below.

UTP = _____   LTP = _____

5.   Slowly decrease the amplitude of the input signal until the output no longer changes states. Record the peak value of the input signal at which this occurs.

V = _____

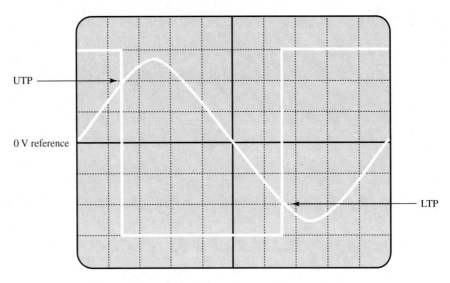

FIGURE 45.3   Measuring the UTP and LTP.

Part 2: Asymmetrical Trigger Points

6. Refer to the circuit shown in Figure 45.4. Calculate the UTP and LTP values for this circuit, and record these values below.

UTP = _____   LTP = _____

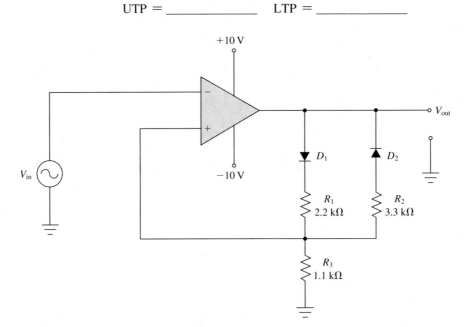

FIGURE 45.4   An asymmetrical inverting Schmitt trigger.

7. Construct the circuit shown in Figure 45.4. Apply a 10 V_{PP} sine wave at a frequency of 1 kHz to the input. As you did in Step 3, observe both waveforms, and draw them in Figure 45.5.
8. Use the input and output waveforms to determine the UTP and LTP values of this circuit, as illustrated in Figure 45.3. Record these values below.

UTP = _____   LTP = _____

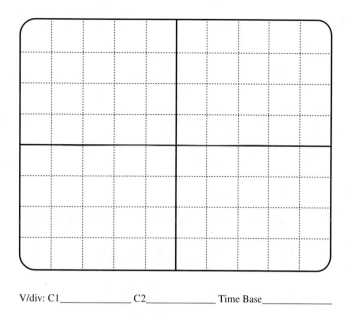

V/div: C1_____ C2_____ Time Base_____

FIGURE 45.5   Waveforms for Step 7.

9.   Disconnect power from the circuit, change R_1 to 10 kΩ, and restore power. Draw the input and output waveforms in Figure 45.6.

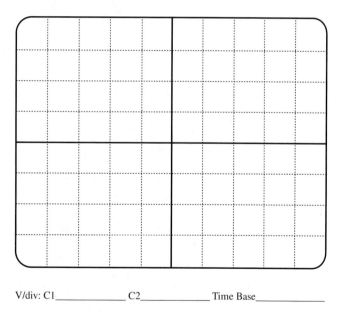

V/div: C1_____ C2_____ Time Base_____

FIGURE 45.6   Waveforms for Step 9.

10.   Measure the UTP and LTP, and record these values below.

UTP = _____ LTP = _____

QUESTIONS & PROBLEMS

1. Refer to Steps 1 and 4. How close were your measured and calculated UTP and LTP values? Explain any significant discrepancies.

2. Refer to Step 5. Explain why the circuit output stopped changing state at this input voltage.

3. Refer to Steps 6 and 8. How close were your measured and your calculated UTP and LTP values? Explain any significant discrepancies.

4. Refer to Steps 9 and 10. Which trigger point was affected by increasing the value of R_1? Explain why this occurred.

Procedure

Open file Ex45.1 from the Electronics Technology Fundamentals companion web site (www.prenhall.com/paynter). This file contains the circuits from Figures 45.1 and 45.4. Run the simulation and use the oscilloscope to verify that the circuits are operating normally. Check the frequency and wave shapes, and note the amplitudes of the input and output waveforms. Once you have verified that your circuits are operating normally, proceed with the fault simulations portion of the exercise.

> *Note:* Some faults that are inserted using the simulator can be dangerous to equipment and/or personnel when they occur in real circuits. Simulations provide exposure to such faults in a safe environment.

Fault Simulations

Fault 45-1	Open R_1 (Symmetrical Schmitt trigger).
Fault 45-2	Shorted R_1 (Symmetrical Schmitt trigger).
Fault 45-3	Open R_2 (Symmetrical Schmitt trigger).
Fault 45-4	Shorted R_2 (Symmetrical Schmitt trigger).
Fault 45-5	Open D_1 (Asymmetrical Schmitt trigger).
Fault 45-6	Shorted D_1 (Asymmetrical Schmitt trigger).
Fault 45-7	Open R_3 (Asymmetrical Schmitt trigger).

Exercise 46

The 555 Timer

OBJECTIVES

After completing this exercise, you should be able to:

- Analyze the basic operation of a 555 timer one-shot.
- Determine the pulse duration of a 555 timer one-shot.
- Analyze the basic operation of a 555 timer free-running multivibrator.
- Determine the operating frequency and duty cycle of a 555 timer free-running multivibrator.

DISCUSSION

The 555 timer contains all the active components needed to construct a *one-shot* (monostable multivibrator) or a *free-running* (astable) multivibrator. All you need to construct one of these circuits is a 555 timer, a few passive components, and a power source.

A 555 timer one-shot is shown in Figure 46.1. When a valid trigger signal is provided at pin 2, the output from the timer (pin 3) goes high for a length of time that is determined by R_A and C_1. At the end of this time, the output automatically returns to its low voltage level and remains there until another trigger signal is received. In the first part of this exercise, you will look at the basic characteristics of the 555 timer one-shot.

The 555 timer free-running multivibrator produces a rectangular-wave output with a duty cycle that can be varied from approximately 50% to approximately 100%. The timing resistors, R_A and R_B, determine the duty cycle. The timing resistors, along with a timing capacitor (C_1), control the oscillating frequency. For duty cycles of greater than 50%, R_A must be greater than R_B. For a duty cycle near 50%, R_A must be lower than R_B. An exact 50% duty cycle, or a duty cycle of less than 50%, can only be achieved with the addition of an external *steering diode*. In the second part of this exercise, you will learn how to vary the timing components to produce an output with a duty cycle of greater than 50%, near 50%, and less than 50%.

LAB PREPARATION

Review Section 24.4 of *Electronics Technology Fundamentals*.

MATERIALS

1   variable dc power supply
1   dual-trace oscilloscope
1   DMM
1   protoboard
6   resistors: 1 kΩ, 10 kΩ, 33 kΩ, 47 kΩ, 220 kΩ, and 470 kΩ
4   capacitors: 0.01 μF, 0.1 μF, 4.7 μF, and 10 μF
1   1N4148 signal diode
1   LM555 timer (or equivalent)
1   momentary push-button switch (NO)

PROCEDURE

Part 1: The 555-Timer One-Shot

1. Calculate the duration of the output pulse for the one-shot shown in Figure 46.1, and record this value below.

$$PW = \text{_____}$$

2. Construct the circuit shown in Figure 46.1. Use your DMM to measure the voltage at pins 2 and 3, and record these values below.

$$V_T(\text{pin 2}) = \text{_____} \quad V_{\text{out}}(\text{pin 3}) = \text{_____}$$

3. Leave your DMM connected to pin 3, and push the momentary switch to trigger the one-shot. Using a watch or stopwatch, estimate the output pulse duration. (*Note:* You may want to do this several times and determine the average pulse duration.) Record this value below.

$$PW = \text{_____}$$

4. Disconnect power from the circuit, and replace R_A with a 220 kΩ resistor. Restore power, and repeat Step 3.

$$PW = \text{_____}$$

5. Disconnect power from the circuit, and restore R_A to its original 470 kΩ value. Change C_1 to 10 μF, restore power, and repeat Step 3.

$$PW = \text{_____}$$

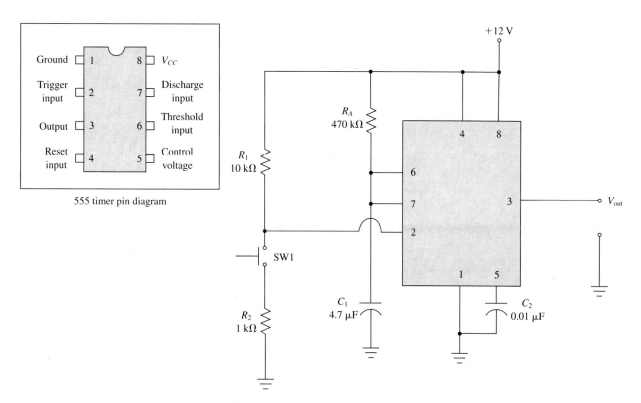

FIGURE 46.1   A 555 timer one-shot.

Part 2: The 555-Timer Free-Running Multivibrator

6. Calculate the output frequency and duty cycle for the free-running multivibrator shown in Figure 46.2, and record these values below.

$f_0 = $ _____ duty cycle $= $ _____

7. Construct the circuit shown in Figure 46.2. Connect your oscilloscope to display the pin 6 (threshold) and pin 3 (output) waveforms. Draw these waveforms to scale in Figure 46.3.

8. Measure the period and pulse width of the output signal (pin 3). Use these measured values to calculate the output frequency and duty cycle, and record all these values below.

$T = $ _____ $PW = $ _____

$f_0 = $ _____ duty cycle $= $ _____

9. Disconnect power from the circuit, and replace R_A with a 47 kΩ resistor. Restore power, and repeat the measurements and calculations in Step 8.

$T = $ _____ $PW = $ _____

$f_0 = $ _____ duty cycle $= $ _____

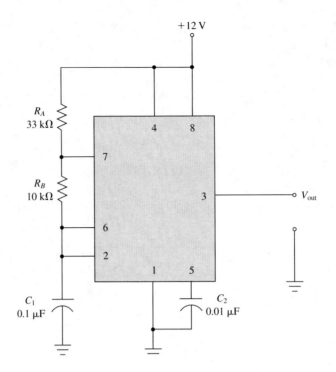

FIGURE 46.2   A free-running multivibrator.

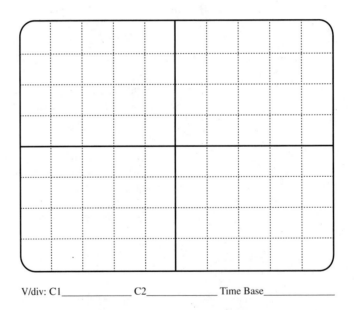

V/div: C1_____ C2_____ Time Base_____

FIGURE 46.3   Waveforms for Step 7.

10.   Disconnect power from the circuit, and replace R_A with a 1 kΩ resistor. Restore power, and repeat the measurements and calculations in Step 8.

$T =$ _____   PW = _____

$f_0 =$ _____   duty cycle = _____

11. Connect the 1N4148 signal diode across R_B as shown in Figure 46.4. As you did in Step 7, measure the waveforms at pins 3 and 6, and draw these waveforms to scale in Figure 46.5.

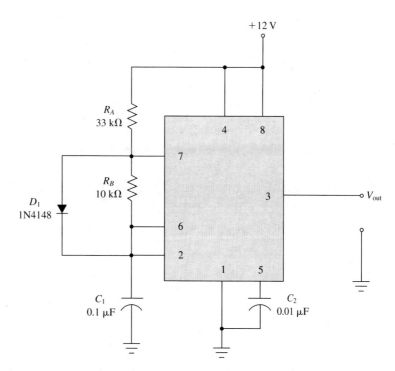

FIGURE 46.4 Modified free-running multivibrator.

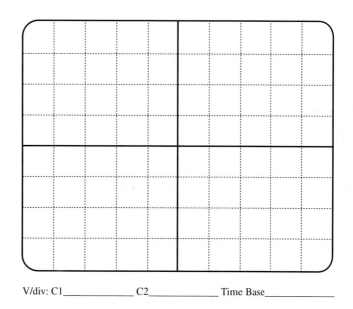

V/div: C1_____ C2_____ Time Base_____

FIGURE 46.5 Waveforms for Step 11.

12. Measure the period and pulse width of the output signal (pin 3). Use these values to calculate the circuit operating frequency and duty cycle, and record all these values below.

$T =$ _____ PW = _____

$f_0 =$ _____ duty cycle = _____

QUESTIONS & PROBLEMS

1. Refer to Steps 1 and 3. How did your measured values of PW compare with the calculated values? Explain any discrepancies between the two values.

2. Refer to Step 2. What state was pin 2 in before the device was triggered? Based on the internal structure of the 555 timer, explain why pin 2 was initially in this state.

3. Refer to Steps 4 and 5. Based on your results, explain what effect R_A and C_1 have on the pulse duration of the one-shot circuit.

4. Refer to Steps 8 through 10. Based on these results, explain what effect changing R_A has on the output frequency.

5. Refer to Steps 11 and 12. In these steps, you should have observed an output with a duty cycle of less than 50%. Explain how the addition of the diode provided a duty cycle that was less than 50%.

SIMULATION EXERCISE

Procedure

1. Open file Ex46.1 from the Electronics Technology Fundamentals companion web site (www.prenhall.com/paynter). This file contains a 555 timer one-shot like the one in Figure 46.1. Note the component value changes to R_A and C_1. You should also note that R_2 has been replaced by a 10-kΩ potentiometer, initially set to 1 kΩ.
2. Run the simulation and use the oscilloscope to verify that the output of the one-shot is changing states when the momentary switch is closed. (V_{out} should change from 0 V to 12 V.)
3. Increase the value of R_2 until the output no longer changes state when the switch is closed. Determine this value of R_2 and record it below.

$$R_2 = \underline{\hspace{2in}}$$

4. Change the value of V_{CC} to 8 V and repeat Step 3.

$$R_2 = \underline{\hspace{2in}}$$

5. Open file Ex46.2 from the Electronics Technology Fundamentals companion web site (www.prenhall.com/paynter). This file contains a free-running multivibrator like the one in Figure 46.2. Note the component value changes to R_A, R_B, and C_1.

6. Run the simulation and determine the minimum and maximum values of the voltage at the threshold input (pin 6). Also, measure the output frequency and amplitude. Record all of these values below.

$$V_{min} = \underline{\hspace{3cm}} \qquad f_0 = \underline{\hspace{3cm}}$$

$$V_{max} = \underline{\hspace{3cm}} \qquad V_{out} = \underline{\hspace{3cm}}$$

7. Change the supply voltage from 12 V to 15 V and repeat Step 6.

$$V_{min} = \underline{\hspace{3cm}} \qquad f_0 = \underline{\hspace{3cm}}$$

$$V_{max} = \underline{\hspace{3cm}} \qquad V_{out} = \underline{\hspace{3cm}}$$

8. Connect a 100 Ω load to the output (pin 3) and measure the output voltage. Change the load to 10 Ω and repeat. Record both of these values below.

$$V_{out} = \underline{\hspace{3cm}} \qquad (R_L = 100\ \Omega)$$

$$V_{out} = \underline{\hspace{3cm}} \qquad (R_L = 10\ \Omega)$$

QUESTIONS

1. Refer to your results from Steps 3 and 4. Explain why the output would not change state when R_2 reached this value. Why did this occur at the same value of R_2 despite the change in V_{CC}?

2. Refer to your results from Steps 6 and 7. Comment on any change, or lack of change, that occurred as a result of a change in V_{CC}.

3. Refer to your results from Step 8. What do they tell you about the 555 timer?

Exercise 47

IC Voltage Regulators

OBJECTIVES

After completing this exercise, you should be able to:

- Analyze the operation of the LM317 voltage regulator.
- Determine the line and load regulation of a voltage regulator.
- Demonstrate the effect that filtering has on the *ripple rejection ratio* of a voltage regulator circuit.
- Build a simple LM317 regulated dc power supply.

DISCUSSION

IC voltage regulators have replaced discrete voltage regulators in most practical linear power-supply applications for several reasons:

- The improved line and load regulation capabilities of IC regulators.
- The decreased size and cost of IC regulators.

In the first part of this exercise, you will learn how to set the output voltage of the LM317. You will also look at the line and load regulation capabilities of the component.

Nearly all voltage regulators attenuate the ripple in the output voltage from a filtered rectifier. This ability is expressed as the *ripple rejection ratio* of the IC regulator. For example, the LM317 has a 65 dB ripple rejection ratio, meaning that its output ripple is 65 dB lower than its input ripple. In the second part of this exercise, you will work with a basic power supply that is regulated by an LM317. As you will see, this ripple rejection of the component can be improved further with the addition of an output capacitor.

Review:

- Sections 25.1 and 25.4 of *Electronics Technology Fundamentals*.
- The specification sheet for the LM317 linear IC voltage regulator.

MATERIALS

1   variable dc power supply

1   dual-trace oscilloscope

2   DMMs

1   transformer (25.2 V secondary)

4   1N4001 rectifier diodes

1   LM317 variable voltage regulator (T0-220 case)

1   heat sink for a T0-220 case

4   resistors: 100 Ω, 220 Ω, 1 kΩ, and 100 kΩ

1   wire-wound resistor: 500 Ω (5 W rated)

1   5 kΩ potentiometer

4   capacitors: 0.1 μF, 10 μF (2), and 470 μF (50 V_{dc} rated)

1   ½ A fuse and fuse holder

PROCEDURE

Part 1: The LM317 Voltage Regulator

1. Construct the circuit shown in Figure 47.1. Connect one DMM to measure the regulator output voltage and the other DMM to measure the input voltage. Initially, use a 1 kΩ load.

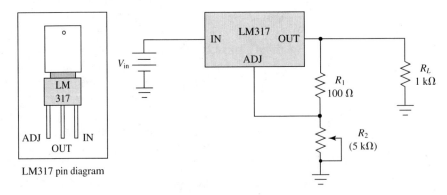

FIGURE 47.1   LM317 test circuit.

2. Set V_{in} to 35 V. Adjust R_2 between its two extremes, and determine the range of V_{out}. Record these values below.

$$V_{out} = \underline{\hspace{2cm}} \text{ to } \underline{\hspace{2cm}}$$

3. Adjust R_2 for a 12 V output.

4. Without disturbing the setting of R_2, record the output from the regulator for values of $V_{in} = 15$ V and $V_{in} = 25$ V. Record the output voltages below.

$$V_{out} = \underline{\hspace{3cm}} @ V_{in} = 15 \text{ V}$$

$$V_{out} = \underline{\hspace{3cm}} @ V_{in} = 25 \text{ V}$$

5. Use your results from Step 4 to calculate the line regulation of the circuit shown in Figure 47.1. Record your answer below.

$$\text{line regulation} = \frac{\Delta V_{out}}{\Delta V_{in}} = \underline{\hspace{3cm}}$$

6. Set V_{in} to 10 V, and adjust R_2 so that $V_{out} = 4$ V.
7. Disconnect power from the circuit. Change R_L to 100 Ω, and restore power.
8. Measure V_L, and use this value to calculate I_L. Assume that 100 Ω is the *full-load* condition of this circuit, and record the values of V_L and I_L in Table 47.1.

TABLE 47.1 Load Regulation

R_L	V_L (Measured)	I_L (Calculated)
100 Ω (full-load)		
100 kΩ (no-load)		

9. Disconnect power from the circuit, and change R_L to 100 kΩ. Assume this is the *no-load* state.
10. Restore power, measure V_L, and calculate I_L. Enter your results in Table 47.1.
11. Using the results in Table 47.1, calculate the load regulation of the circuit in Figure 47.1. Record your answer below.

$$\text{load regulation} = \frac{V_{NL} - V_{FL}}{\Delta I_L} = \underline{\hspace{3cm}}$$

12. Disconnect power from the circuit, and remove the potentiometer. Use your DMM to measure its minimum and maximum values. (You will need these values for the questions at the end of this procedure.) Record your answers below.

$$R_{min} = \underline{\hspace{2.5cm}} R_{max} = \underline{\hspace{2.5cm}}$$

Part 2: A Practical DC Power Supply (Optional)

13. Construct the circuit shown in Figure 47.2. Connect your DMM across the load, and adjust R_2 for an output of 15 V.
14. Connect your oscilloscope so that Channel 1 displays the ripple voltage at the regulator input and Channel 2 displays the regulator output ripple voltage. Both channels must be ac coupled. Measure these two values, and record them below.

$$V_{r(in)} = \underline{\hspace{2.5cm}} V_{r(out)} = \underline{\hspace{2.5cm}}$$

15. Disconnect power from the circuit, and remove the output filter capacitor (C_4). Restore power, and repeat Step 14. Record your measurements below.

$$V_{r(in)} = \underline{\hspace{2.5cm}} V_{r(out)} = \underline{\hspace{2.5cm}}$$

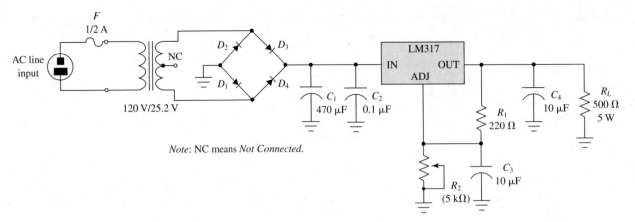

FIGURE 47.2 A regulated dc power supply.

Note: NC means *Not Connected*.

QUESTIONS & PROBLEMS

1. Refer to Steps 2 and 12. Using the minimum and maximum resistance of the 5 kΩ pot, calculate the expected output voltage using:

$$V_{out} \cong 1.25 \left(\frac{R_2}{R_1} + 1 \right) = \underline{\hspace{2cm}} \text{ (minimum)}$$

$$V_{out} \cong 1.25 \left(\frac{R_2}{R_1} + 1 \right) = \underline{\hspace{2cm}} \text{ (maximum)}$$

Were the minimum and maximum output voltages close to the values that you calculated? Explain any significant discrepancies.

2. Refer to Steps 5 and 11 and the spec sheet for the LM317. Were the load and line regulation values that you calculated within the specified range for this device? If not, explain any discrepancies.

3. Refer to Step 14. Why did both channels of the scope need to be ac coupled?

4. Refer to Steps 14 and 15. Calculate the ripple rejection ratio for each one of these conditions.

$$\text{rejection ratio} = \underline{\hspace{2cm}} \ (C_4 \ \text{installed})$$
$$\text{rejection ratio} = \underline{\hspace{2cm}} \ (C_4 \ \text{removed})$$

a. Did the ratio change from condition to condition?

b. Which condition had the best ratio? Explain why you think this was the case.

Exercise 48

Silicon-Controlled Rectifiers

OBJECTIVES

After completing this exercise, you should be able to:

- Demonstrate the latching action of a silicon-controlled rectifier (SCR).
- Analyze the operation of an SCR phase-control circuit.

DISCUSSION

Silicon-controlled rectifiers, or SCRs, belong to a group of components called *thyristors*. SCRs are three-terminal, four-layer devices that use internal feedback to produce a *latching* (breakover) action. Introduced in 1956 by Bell Labs, SCRs are *unidirectional* devices; that is, they conduct in one direction only.

An SCR is commonly triggered by a gate input signal. Once it *fires* (or *latches*) into conduction, an SCR will not respond to another gate trigger signal. The only way that the SCR can be turned off is by reducing its anode current below the required *holding current*. In the first part of this exercise, you will investigate these basic characteristics of the SCR.

In the second part, you will construct and analyze a common SCR circuit. A *phase controller* is a circuit that controls load power by controlling the *conduction angle* through that load; that is, it controls the portion of the input waveform that is coupled to the load. Since SCRs are unidirectional, the maximum conduction angle is 180° of the input signal.

Review:

- Sections 26.1 and 26.2 of *Electronics Technology Fundamentals*.
- The specification sheet for the 2N5060 SCR.

MATERIALS

1	variable dc power supply
1	dual-trace oscilloscope
2	DMMs
1	function generator
1	protoboard
1	power transformer: 25.2 V
1	1N4001 rectifier diode
1	LED
1	2N5060 SCR
2	resistors: 1 kΩ
2	potentiometers: 10 kΩ and 100 kΩ
1	0.01 μF capacitor
1	¼ A fuse and fuse holder

PROCEDURE

Part 1: Latching Mode

1. Construct the circuit shown in Figure 48.1. Note that the DMMs are connected as ammeters. The 100 kΩ pot should be set initially to its maximum value and the 10 kΩ pot to its minimum value.

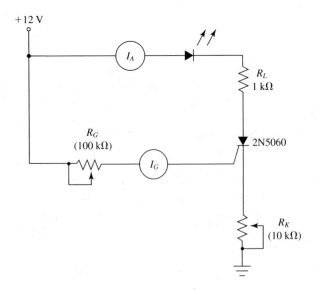

FIGURE 48.1 An SCR test circuit.

2. Apply power to the circuit. The LED should remain off. Decrease the R_G setting *slowly* until the SCR fires (latches) and the LED lights. *(Note: Adjust R_G only as far as required to light the LED.)*
3. Measure the gate current. This is the *gate-trigger current* (I_{GT}). Then, measure the voltage across the gate-cathode junction. This is the *gate-trigger voltage* (V_{GT}). Record both these values below.

$$I_{GT} = \underline{\hspace{3cm}} \qquad V_{GT} = \underline{\hspace{3cm}}$$

4. Now, slowly increase R_G until it is back at its maximum value. Then, without powering down, carefully disconnect the gate lead to the SCR. Does the LED turn off under any of these conditions?

5. Reconnect the gate lead and slowly increase R_K while closely monitoring the current in the anode (I_A). A point will be reached when the SCR stops conducting. The anode current just before this happens is the *holding current* (I_H). Record your estimate of the holding current below. (*Note:* You may want to repeat this process a few times to get an accurate value.)

$$I_H = \underline{\hspace{3cm}}$$

Part 2: An SCR Phase Controller

6. Construct the circuit shown in Figure 48.2. Initially, the 10 kΩ pot should be set to approximately halfway between its extremes. Connect Channel 1 of your scope to display the input signal and Channel 2 to display the load voltage. Both channels should be dc coupled.
7. Observe the two waveforms, and draw them to scale in Figure 48.3.
8. Carefully adjust R_2, and determine the minimum and maximum firing angles of this phase-controller circuit. (*Note:* The firing angle is the point on the input waveform, in degrees, where the SCR fires.)

$$\text{firing angle} = \underline{\hspace{3cm}} \text{(minimum)}$$
$$\text{firing angle} = \underline{\hspace{3cm}} \text{(maximum)}$$

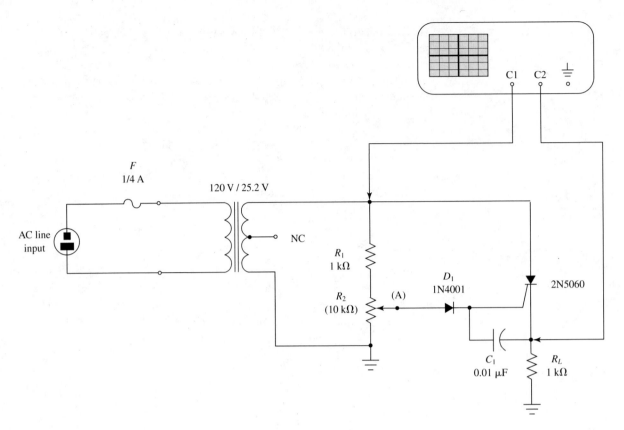

FIGURE 48.2 An SCR phase controller.

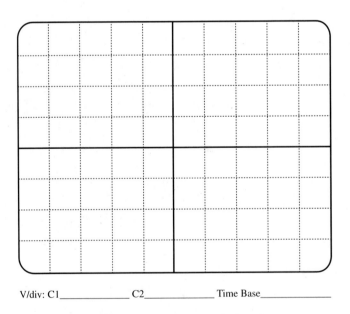

V/div: C1_____ C2_____ Time Base_____

FIGURE 48.3 Waveforms for Step 7.

1. Refer to the specification sheet for the 2N5060 SCR. Determine the rated values for I_{GT}, V_{GT}, and I_H, and record your answers below.

$$I_{GT} = \text{_____} \text{ (rated)}$$

$$V_{GT} = \text{_____} \text{ (rated)}$$

$$I_H = \text{_____} \text{ (rated)}$$

Now, refer to the values that you measured in Steps 3 and 5. How did your measured values compare with the rated values? Identify and explain any discrepancies.

2. Refer to Step 4. Why did the LED remain lit even with the gate disconnected?

3. Refer to the waveforms that you recorded in Figure 48.3. Explain the output waveform in terms of circuit operation.

4. Refer to Figure 48.2. What function does C_1 play in this circuit?

Appendix A

Circuit Construction and the Protoboard

PROTOBOARDS

The *protoboard* is a plug-in, solderless base that allows you to construct and test simple circuits without having to solder the components together. The protoboard is also commonly referred to as a *breadboard* or a *prototyping board*.

The layout of a typical protoboard is illustrated in Figure A.1. The protoboard is divided in half horizontally by a gap. The upper and lower portions of the board are mirror images of each other. Across the top and the bottom of the protoboard are two horizontal rows of plug-in sockets. All the sockets in each row are connected internally to each other, as shown in Figure A.2. This makes an electrical connection between any components or wires inserted into the same row. Note, however, that each horizontal row is electrically isolated from the row below or above. In other words, there is no continuity between rows, only between sockets in the row. These horizontal rows are usually used for voltage supply or ground connections, as shown in Figure A.3.

Between each pair of horizontal rows and the central gap are a number of vertical columns. There are five internally connected sockets per column, as shown in Figure A.2. If one or more wires or component leads are inserted into any of these five sockets, they make an electrical connection as if they were soldered together. There is no continuity between columns, only between sockets in the column.

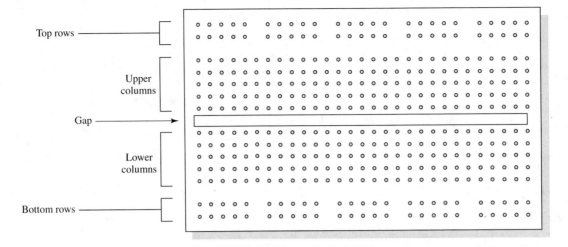

FIGURE A.1 Typical protoboard layout.

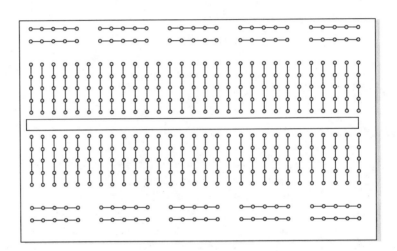

FIGURE A.2 Protoboard internal connections.

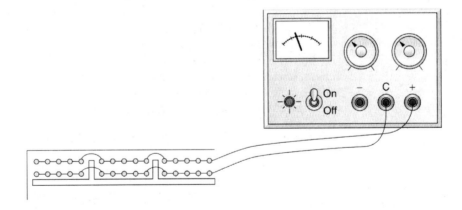

FIGURE A.3 Typical dc power-supply connections.

As with any other skill, certain rules and guidelines will help you to construct a circuit using the protoboard that is both accurate and easy to analyze. Here are a few do's and don'ts to keep in mind:

- Use only 22 or 24 gauge solid-core wire (jumper wires) to connect components. Higher-gauge wires may be too thin to make a solid mechanical connection in the socket and, thus, may make a poor or intermittent electrical connection. Lower-gauge wires may be too thick for the sockets, and thus, may damage the socket when inserted.
- Use ¼ or ½ W resistors only. As with lower-gauge wires, the leads of a higher-wattage resistor may be too large for the plug-in sockets. (If an exercise calls for use of a higher-wattage resistor, use connecting wires and/or alligator clips to connect it to the circuit on the board.) At the same time, the leads of a lower-wattage resistor may be too thin to make a solid mechanical (and electrical) connection to the protoboard.
- Try to build a given circuit so that it resembles the schematic as much as possible. This idea is illustrated in Figure A.4. The circuit shown in Figure A.4b looks very much like the schematic shown in Figure A.4a. If you should encounter a problem when analyzing a circuit, either you or your instructor may need to *troubleshoot* that circuit. If it is laid out like the one shown in Figure A.4b, any construction fault will be relatively easy to find. If it is laid out like the circuit shown in Figure A.4c, the fault will be much harder to locate. Neither you nor your instructor can easily tell which connections are correct and which are not.

The circuit shown in Figure A.4 is relatively simple. Some of the circuits that you will have to construct, however, will be more complex. As the complexity of the circuit increases, so does the potential for construction error. This means that a neatly constructed circuit becomes more important as the circuit complexity increases.

- Keep your jumper wires as short as is practical, and keep all wires and components tight to the board. Long jumper wires make for a sloppy board. In a tangle of wires, it is very difficult to tell if one should come loose or has been mistakenly inserted into the wrong socket. The same is true of the components. If the components are kept tight to the board, the circuit is neater and easier to troubleshoot.
- Finally, a word of caution. Never mount a component with both of its leads inserted in the same column, as shown in Figure A.5a. Remember that all the sockets in any given column are connected together internally. This means that if you mount the component as shown in Figure A.5a, the two component leads will be connected, which will effectively short out the component. Depending on the location of the component in the circuit, this can result in an overload condition that may damage one or more of the other components in the circuit. In contrast, when connected as shown in Figure A.5b, the current must pass through the component.

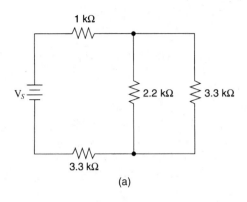

(a)

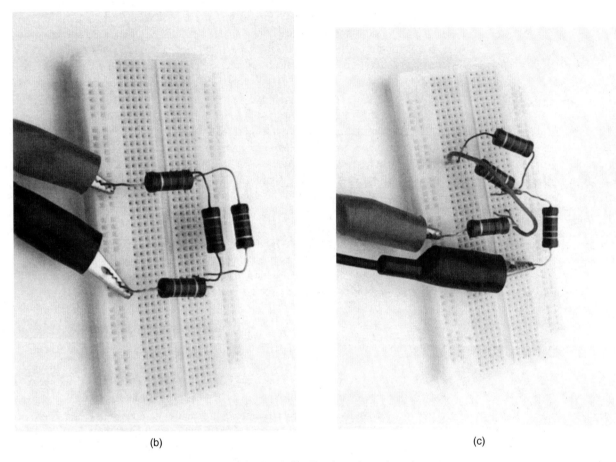

(b)   (c)

FIGURE A.4   Circuit construction.

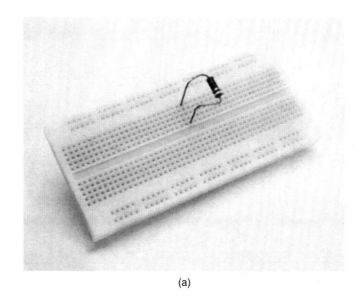

(a)

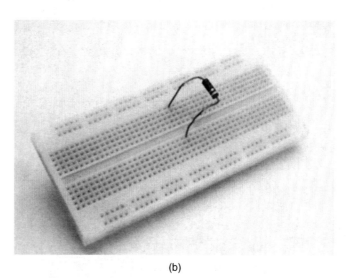

(b)

FIGURE A.5 Component connections on a protoboard.

Appendix B

Multimeters

Multimeters can be divided into two distinct categories: digital meters and analog meters. Digital multimeters, or DMMs, are generally preferred for a variety of reasons that we will discuss later. The analog meter is usually referred to as a VOM, which is an acronym for *volt-ohm-milliammeter*.

A DMM and a VOM are shown in Figure B.1. The DMM processes its measurement internally and produces a digital readout. The VOM uses a current-driven indicator needle that moves across a fixed scale. The position of the needle, with respect to the scale, indicates the value of the measurement. Since DMMs are more commonly used, we will concentrate on them in this discussion. Even so, VOMs are still in use, and as a trained technician, you should be capable of using either instrument correctly.

As its name implies, the multimeter can be used to take a variety of measurements. At a minimum, it can be used to measure resistance, ac and dc voltage, and dc current. Some multimeters offer other functions as well, such as ac current measurement, a continuity checker, a diode checker, and a transistor current gain (h_{FE}) checker. Some models even include a frequency counter and capacitance meter. In this section, we will limit our discussion to the following measurements:

- Resistance measurements.
- Voltage measurements.
- Current measurements.
- Diode testing.
- Transistor testing.

FIGURE B.1 A DMM and a VOM. *(Figure B.1a provided by Altadox, Inc.*
Figure B.1b provided by Simpson Electric Co.)

AUTO-RANGING METERS

Many meters require the user to set the range. Setting the meter to the correct range is necessary to get the most precise reading and, in some cases, to protect the meter from damage. Other meters are *auto-ranging,* meaning that they automatically set themselves to the optimum measurement range for the value being measured.

MEASURING RESISTANCE

When measuring resistance, two rules must *always* be followed:

- Make sure that the circuit power is off. Never make a resistance measurement with the circuit powered up.
- Isolate the component that you are measuring. This can be done by completely removing it from the circuit or simply by disconnecting one lead. (The reason for this will become clear when you study parallel circuits.)

Once you have removed power and isolated the component(s):

- Set the function switch of your DMM to the "ohms" position.
- Set the range to one that is just higher than the resistance you want to measure. For example, suppose that you want to measure a resistor with a rated value of 4.7 kΩ.

If your meter has range settings of 200 Ω, 2 kΩ, 20 kΩ, 200 kΩ, and 2 MΩ, you should choose the 20 kΩ range. (If you do not know the resistance value that you are measuring, set the meter to its highest range.) After the meter is connected to the component, decrease the range setting until you reach the lowest setting that does not result in an "out of range" reading. (Check the operation manual to determine how your meter displays an out of range condition.)

• Connect your test leads. All meters have a common (usually labeled as "COM") lead socket. Connect your black lead to this socket. The red lead should be connected to the socket that is labeled as VΩ or VΩA. (Check the operation manual of your meter if you are in doubt.)

Now you are ready to make a resistance measurement.

When making a resistance measurement, touch only one lead of the resistor with your fingers, especially if you are making high-value resistance measurements. The reason for this is that the human body has its own resistance. If both leads of the resistor come in contact with your body, you can affect the measurement. The proper technique for making a resistance measurement is illustrated in Figure B.2.

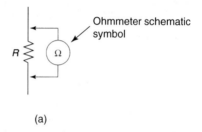

(a)

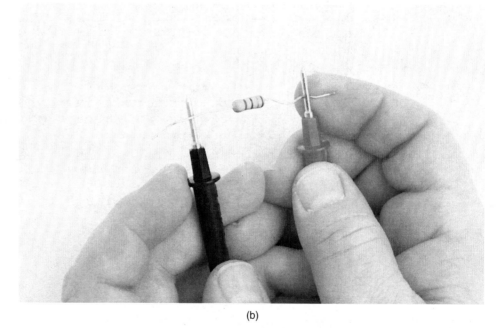

(b)

FIGURE B.2 Measuring resistance.

VOM RESISTANCE MEASUREMENTS

VOM resistance measurements are a little more complicated than DMM measurements. For one thing, each time you change ranges, you must "zero" the meter. To zero a VOM:

- Select the desired range, and then short the leads together. The needle will deflect across the scale to the right-hand side.
- With the leads shorted, use the *zero adjust* control to move the needle so that it is directly over the 0 Ω position on the scale.

These two steps eliminate factors that can affect the meter resistance reading, such as the resistance of the leads themselves.

VOMs provide a readout via an indicator needle above a stationary scale. As such, they are subject to what is called *parallax error*. If you look at the meter from one side or the other, the space between the needle and the scale can cause you to get two different readings—even though the indicator needle hasn't moved. (Try it!) To get an accurate reading, *look directly down on the scale*. Many meters include a reflective strip just below or above the scale that can eliminate parallax error. If you adjust your viewing angle until the needle covers its own reflection, you are viewing it from directly above, and parallax error is avoided.

Setting the range correctly is even more important when using a VOM. If you look at the ohms scale you will see that the divisions get closer together—that is, the resolution gets poorer—the farther left that you go on the scale. This is because it is a logarithmic scale. To get the most accurate reading, try to set the meter range so that the indicator needle rests in the right-hand half of the scale when the measurement is made.

MEASURING VOLTAGE

Voltage is measured by connecting the DMM *across* the component(s) being tested, as illustrated in Figure B.3. As you can see, the meter leads are connected on the two sides of the component. Note that the circuit current path is not broken. As with measuring resistance, the black meter lead is connected to the common (COM) socket on the meter, and the red lead is connected to the VΩ or VΩA socket.

To measure the voltage across a component (or group of components):

- Determine whether you want to make a dc or an ac voltage measurement, and set the function switch accordingly. (Check the operation manual of your meter if you aren't sure how to set the meter function.)
- Set the range of the meter to the correct value. The same criteria are used as in resistance measurement. You want the lowest range that does not result in an out-of-range reading to get the most accurate measurement. If you are making an ac voltage measurement, remember that a DMM always gives an rms value, not an average, a peak, or a peak-to-peak value.

For any dc voltage measurement, you will read a *positive* or a *negative* value. If the red (positive) lead is connected to the side of the component that is more positive, your reading will be positive. If the black (common) lead is connected to the more positive side, your reading will be negative. With a DMM, it is not important which way the leads are connected as long as you understand what the reading is telling you. Polarity is not an issue when making an ac voltage measurement, since the polarity of the voltage is constantly changing.

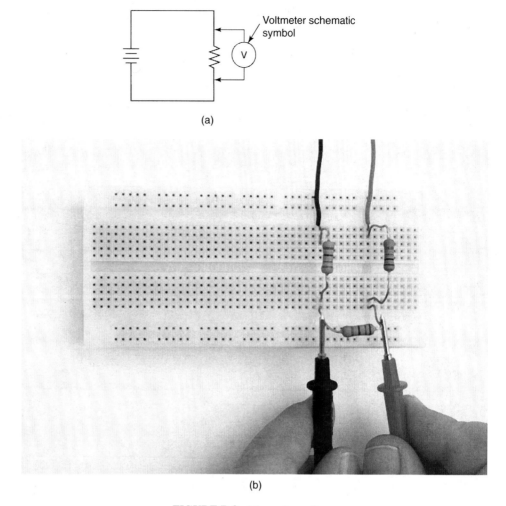

(a)

(b)

FIGURE B.3 Measuring voltage.

VOM VOLTAGE MEASUREMENTS

To make a voltage measurement with a VOM, you must be more careful about polarity and range, because the meter movement of a VOM is quite delicate and can be easily damaged by "pegging" the needle. If the range you choose is too low, the needle will be driven quickly across the scale (left to right) and *pegged* on the high side of the scale. This can bend the needle or burn out the movement. For this reason, always start a voltage measurement on the highest scale.

VOMs cannot adjust automatically for the polarity of a dc voltage. If the meter is connected to the circuit so that the negative (COM) lead is connected to the more positive side of the component, the needle will try to deflect below zero. If you are using a VOM and are unsure about the polarity of the voltage being measured:

- Set the meter range on the highest setting, and make your voltage measurement.
- If the needle deflects to the right, you know that the polarity is correct.
- If the needle deflects to the left (below zero), it should not damage the meter movement when the meter is on its highest scale. You can then reverse the leads (or change the function setting) and adjust the range for the most accurate reading.

Except for these restrictions, the approach to measuring voltage with a VOM is the same as it is for a DMM. It is not necessary to zero the VOM for voltage measurements. However, remember to watch for parallax errors!

MEASURING CURRENT

As mentioned earlier, the current path of the circuit is not broken when measuring voltage. Obviously, this technique will not work if we want to measure current. To measure current, the meter must be connected so that circuit current passes *through* the meter. For the current to pass through the meter, the circuit current path must be broken and then re-established using the meter, as illustrated in Figure B.4.

As usual, the black lead is connected to the common (COM) socket of the meter. The meter connection for the red lead depends on the magnitude of the current and the particular meter you are using. Some meters have a separate socket for low-value current meas-

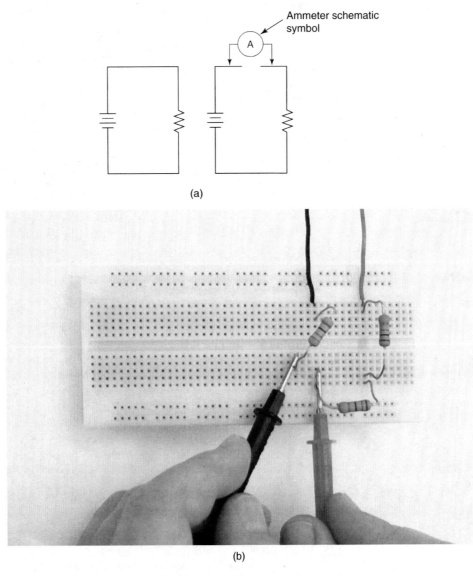

(a)

(b)

FIGURE B.4 Measuring current.

urements (typically labeled as "µA/mA"). As stated earlier, some meters use the same socket for resistance, voltage, and low-value current measurements (typically labeled as "VΩA"). In either case, these sockets are usually limited to a maximum of 300 mA. A second socket is used for high-value current measurements (typically labeled as "10 A" or "20 A," depending on the current-handling capability of the meter). Once again, refer to the operation manual of your meter for clarification.

As with voltage measurements, you must determine if you are making a dc or an ac measurement and then set the function switch of the meter accordingly. Once the type of current is determined, set the meter to the lowest range that does not result in an out of range reading. If the measurement is dc, then polarity is an issue. A negative reading simply means that the leads of the meter are connected backwards. For example, if the positive (red) lead of the meter shown in Figure B.4 is connected to the resistor side of the circuit (right side), then the meter will give a negative reading. When using a DMM, it doesn't matter whether the reading is negative or positive as long as you know what the reading means. When making an ac current measurement, polarity is not an issue, because the polarity of ac current is constantly changing. As with voltage, alternating current (ac) is measured in rms values.

VOM CURRENT MEASUREMENTS

VOMs cannot measure alternating current. To make a dc measurement, the meter must be inserted into the circuit current path (in series), just like the DMM. Polarity and range are more important when working with a VOM for the same reasons mentioned earlier. You must be careful not to damage the meter movement. Always set the meter to the highest range before you power up, and determine if the polarity is correct. If it is, decrease the range to the one that gives you the most accurate reading without going out of range. As with voltage measurements, it is not necessary to zero the meter.

DIODE AND TRANSISTOR TESTING

The resistance of a *forward biased pn*-junction should be relatively low. The resistance of a *reverse biased pn*-junction should be near that of an open circuit. Both DMMs and VOMs can be used to make a resistance check of a diode, as illustrated in Figure B.5. The function switch of the meter is set to measure resistance. To perform a forward resistance check, the positive lead of the meter is connected to the anode of the diode. The negative lead is connected to the cathode. When forward biased, a good diode should have a resistance of less than 1 kΩ.

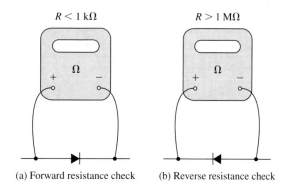

(a) Forward resistance check   (b) Reverse resistance check

FIGURE B.5   Diode testing

To make a reverse resistance check, simply reverse the meter leads. When reverse biased, a good diode should have a resistance of greater than 1 MΩ. If the forward biased resistance is greater than 1 kΩ or the reverse bias resistance is less than 1 MΩ, the diode is suspect and should be replaced.

Transistors have *pn*-junctions just like a diode. To test a BJT you make a resistance check on each pair of terminal junctions, one terminal pair at a time. The secret is to know whether each reading should be a low or a high-resistance reading. Let's assume that you are testing the base-emitter junction of an NPN transistor. If the positive lead is connected to the base, and the common lead is connected to the emitter, the junction is forward biased. If the *pn*-junction is functioning properly, you should get a low-resistance measurement. If you get a *very high* resistance measurement, the junction is most likely open. If you get a *very low* resistance measurement, the junction is most likely shorted. If either of these conditions are measured, the transistor is damaged and should be replaced.

As stated earlier, the secret is in knowing whether you are forward or reverse biasing the junction and thus, what reading you should get when measuring a specific pair of transistor terminals. Table B.1 below lists all of the possible resistance readings you should get for a properly functioning PNP or NPN transistor. Note that there is no *pn*-junction between the collector and emitter terminals. This means that they cannot be forward or reverse biased and should always read as an open circuit.

TABLE B.1

Transistor Type	Junction	Lead Connections	Bias	Resistance
NPN	BE	Red to Base, Black to Emitter	Forward	Low
NPN	BE	Red to Emitter, Black to Base	Reverse	High
NPN	BC	Red to Base, Black to Collector	Reverse	High
NPN	BC	Red to Collector, Black to Base	Forward	Low
NPN	CE	Red to Collector, Black to Emitter	N/A	High
NPN	CE	Red to Emitter, Black to Collector	N/A	High
PNP	BE	Red to Base, Black to Emitter	Reverse	High
PNP	BE	Red to Emitter, Black to Base	Forward	Low
PNP	BC	Red to Base, Black to Collector	Forward	Low
PNP	BC	Red to Collector, Black to Base	Reverse	High
PNP	CE	Red to Collector, Black to Emitter	N/A	High
PNP	CE	Red to Emitter, Black to Collector	N/A	High

Note: The readings in Table B.1 assume that the red meter lead is positive.

Most digital meters have a "diode check" function. Rather than giving a resistance reading, this function measures the voltage drop across the *pn*-junction. The leads of the meter are connected to the diode or transistor terminals in the same way as when making forward and reverse resistance measurements. When forward biased, a normal *pn*-junction diode should read somewhere in the 600 to 800 mV range. Some special application diodes or transistors can read outside of this range. You must check the specification sheet of the device you are testing. When reverse biased, the device should give an out of range reading. Most diode checkers have a limit of 2 V maximum.

As was the case with ohmmeters, the diode check function can be used to test BJT transistors. Refer back to Table B.1. Where the ohmmeter gives a low resistance reading, the diode checker should read between 600 mV and 800 mV. Where the ohmmeter gives a high resistance reading, the diode checker will give an out-of-range reading.

Before we finish, a few notes of caution should be made:

- When the range on some ohmmeters is to set to low resistance scales, they can generate enough current to damage low-current diodes. It is a good rule of thumb to use the R × 1000 scale (or equivalent) when making the forward resistance check.

- Most meters have a negative (COM) terminal, but this is not always the case. You should always check the operations manual of the meter you are using to determine lead polarity. Otherwise, you may think that you have a bad reading when the device is really all right.

- In many cases, diodes and transistors can be tested "in-circuit". The most common failure is a shorted *pn*-junction. If a junction has opened, it is usually as a result of first shorting out, and then burning open. This is a *catastrophic failure* and can often be seen with the naked eye. The problem with testing a device in-circuit is that you often do not know what components are in parallel with it. If a low-resistance component is connected in parallel with the device under test, you might read a very low-value resistance in both directions. If a junction measures as a short when tested in-circuit, remove and retest the component before you decide that it must be replaced.

METER FREQUENCY LIMITS

Both DMMs and VOMs are capable of making ac voltage measurements. Digital meters can also make alternating current measurements, but there is a limit to what they can do. Most multimeters have an upper frequency limit of approximately 10 kHz. Any reading above this frequency is suspect. As usual, check the operating manual of your meter to determine its exact frequency limit. Typically, you must use an *oscilloscope* to measure voltages above 10 kHz. (Oscilloscope measurements are discussed in Appendix C.)

Appendix C

Oscilloscopes

Unlike the multimeter, oscilloscopes are capable of making only one type of measurement. An oscilloscope produces a visual display of *voltage with respect to time*. There are a wide variety of signals whose voltage changes over time; these signals are referred to as *waveforms*. Some of the more common waveforms include:

- Sine waves.
- Square and rectangular waves.
- Triangle and sawtooth waves.
- Step and pulse shapes.

Before proceeding, we should review some of the terminology that is used when discussing waveforms:

- **Period (or cycle time).** The time required to complete one cycle of a repeating waveform.
- **Frequency.** The number of waveform cycles per second.
- **Amplitude.** The magnitude of a waveform.

Period and frequency are intimately related. The frequency of a waveform is equal to the *reciprocal* of its period, and vice versa. By formula,

$$f = \frac{1}{T} \quad \text{and} \quad T = \frac{1}{f}$$

where

T = the period of the waveform
f = frequency of the waveform

Figure C.1 illustrates the concepts of period, peak, and peak-to-peak measurements.

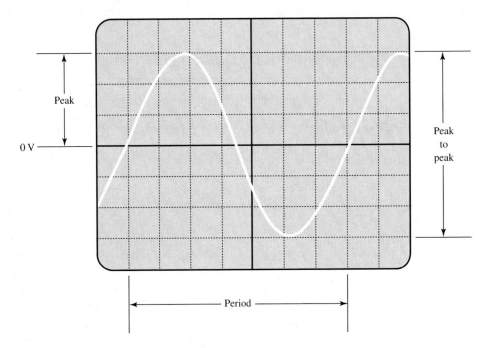

FIGURE C.1   Waveform measurements.

Digital versus Analog Oscilloscopes

Oscilloscopes are available in a range of models, with a wide variety of capabilities and features. Most scopes have at least two inputs (2 channel scopes) but many have more. Despite this variety, they can be broken down into two basic types: the *analog* oscilloscope and the *digital storage* oscilloscope (DSO).

Analog scopes produce a "real time" display of the signal you are measuring. Digital scopes convert the input voltage into a digital format for storage in memory by routing the input through a fast analog-to-digital converter (ADC). They then take samples of the digital information and reconstruct the waveform on the display. As the basic functions are the same for both types of scopes, we will begin by discussing the simpler analog oscilloscope.

FRONT-PANEL CONTROLS

The front panel of a two-channel analog oscilloscope is shown in Figure C.2. In this section, you will look briefly at the four basic types of controls that you will need to know how to use.

The Display

Some analog scopes use liquid crystal displays (LCDs) but many use a cathode ray tube (CRT) as the display medium. A CRT works in the following manner: An electron beam sweeps across the screen at a specific rate, which is controlled by the horizontal deflection

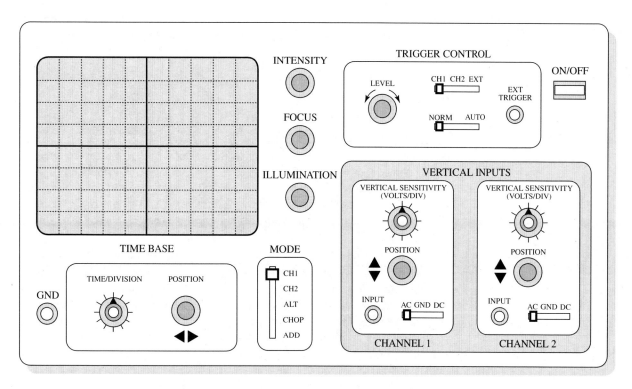

FIGURE C.2   A basic oscilloscope screen and control panel.

circuitry. The inside of the display screen is coated with a *phosphor* that becomes luminous as long as the beam comes in contact with it. This beam striking the phosphor is how the waveform is *"written"* on the display.

The oscilloscope screen displays a two-axis graph, which is commonly called the *grid*. The vertical axis is divided into eight major divisions. The horizontal axis is divided into 10 major divisions. Each major division, on both axes, is divided into five minor divisions.

Display Controls

These controls allow you to adjust various properties of the CRT display. There are three basic display functions:

- **Intensity,** which controls the brightness of the displayed waveform.
- **Focus,** which controls the sharpness and clarity of the displayed waveform.
- **Scale illumination,** which controls the level of back-lighting on the display.

Vertical Deflection Controls

As stated at the beginning of this discussion, an oscilloscope produces a visual display of voltage with respect to time. This is a two-axis display. The *vertical axis* of the graph represents the *amplitude* of the voltage waveform. There are several vertical deflection controls:

- **VOLTS/DIV.** This control, which is often referred to as *vertical sensitivity,* determines the voltage represented by each major division of the vertical axis. Refer back to Figure C.1. As you can see, that waveform occupies exactly six major divisions (peak-to-peak) on the vertical axis. If the VOLTS/DIV control was set to 10 mV/div, this waveform would represent a signal with an amplitude of

$$6 \text{ div} \times 10 \text{ mV/div} = 60 \text{ mV}_{PP}$$

If the control was set to 5 V/div, then the amplitude would be 30 V_{PP}.

- **Variable VOLTS/DIV.** This control allows you to set the vertical scale to a *non-calibrated* value by rotating the knob in the center of the VOLTS/DIV control. When turned completely clockwise, the vertical scale is calibrated to the VOLTS/DIV setting. When rotated counter-clockwise, the vertical scale is no longer calibrated, and the value of the waveform amplitude cannot be determined. This control is used to produce a waveform display of a desired height. (Later in this appendix, you will be shown a use for this control.)
- **Position.** This control moves the waveform up or down on the grid. It is also used to establish a baseline or reference for the waveform trace.
- **Display Mode Select (CH1-CH2-ALT-CHOP-ADD).** This control determines which of the input channels of the scope are shown and the manner in which they are displayed. For a two-channel scope, Channel 1, Channel 2, or both channels can be displayed. In **ALT** mode, the two channels are alternately displayed, one after the other, in very quick succession. In **CHOP** mode, a small portion of each channel's waveform is displayed, one channel at a time, as the trace moves across the screen. In **ADD** mode, the waveforms of the two channels are added together, and the sum of the two waveforms is displayed.
- **Input Mode Select (AC-GND-DC).** This control determines the type of input signal that is *coupled* to the scope. When **AC** is chosen, only ac signals are coupled to the scope, and any dc component of the waveform (offset) is blocked by a series capacitor. When **GND** is chosen, a 0 V signal is shown, and the vertical position control (mentioned earlier) can be used to set the 0 V reference position of the display. In this mode, the display shows only a flat, horizontal line. When **DC** is chosen, both ac and dc voltages are coupled to the scope. This allows you to accurately display a waveform with both ac and dc components, such as a sine wave with a dc offset.

Horizontal Deflection Controls

As mentioned earlier, the horizontal axis of the oscilloscope display represents *time*. The basic horizontal deflection controls are:

- **TIME/DIV.** This control, which is often referred to as the *time base,* determines how much time each major division of the horizontal axis represents. Refer back to Figure C.1. One complete cycle of that waveform occupies exactly seven major divisions of the horizontal axis. If the TIME/DIV control was set to 0.5 ms, then the period of this waveform would be found as

$$8 \text{ div} \times 0.5 \text{ ms/div} = 4 \text{ ms}$$

Since frequency is the reciprocal of the waveform period, the frequency of the waveform displayed in the figure can be found as

$$f = \frac{1}{T} = \frac{1}{4 \text{ ms}} = 250 \text{ cycles per second (cps)}$$

As you know, 250 cps would normally be expressed as 250 Hz. If the TIME/DIV control was set to 2 μs/div, then the displayed waveforms would have values of

$$T = 16 \text{ μs and } f = 62.5 \text{ kHz}$$

Note that as the frequency of the waveform you want to display increases, you must *decrease* the time base setting to display a single waveform cycle.

- **Variable TIME/DIV.** This control allows you to take the horizontal display out of its calibrated state. It should be kept in the fully *clockwise* position for the exercises that you will perform in this laboratory manual.
- **Position.** This control allows you to move the waveform to the left or the right on the grid.

Trigger Controls

The electron beam in the CRT travels across the screen at a rate that is determined by the TIME/DIV control. The trigger circuit allows you to synchronize the sweep of the electron beam with a specific point on the waveform that you want to display. Once a specified *threshold voltage* is reached on the waveform, the sweep circuit is *triggered*. This ensures that the waveform is shown in the same location on the grid for each sweep of the electron beam. The controls in the trigger section are:

- **Trigger Mode Select (NORM-AUTO).** When set in the **AUTO** position, the scope triggers on an internal, preset voltage threshold. When set in the **NORM** (normal) position, the scope triggers on the threshold voltage that the user sets by adjusting the level control.
- **Level Control.** This control allows you to set the trigger threshold to a specific voltage value. It also allows you to determine whether the scope triggers on a positive or a negative voltage value. By choosing a positive or negative voltage threshold, you can set the scope to trigger on a positive- or negative-going transition.
- **Trigger Source Select (CH1-CH2-EXT).** This control allows you to determine whether you want the scope to trigger on the Channel 1 input, the Channel 2 input, or some *external* trigger source. None of the exercises in this laboratory manual require the use of an external trigger source.

Digital Storage Oscilloscopes (DSOs)

Just as DMMs have become far more common that VOMs, the digital storage scope has replaced the analog scope in most labs. DSOs have greater capabilities than their analog brethren. The range of features and capabilities varies widely from brand to brand and model to model. For this reason we will focus on the basics, specifically, how digital scopes differ from analog scopes.

As stated earlier, the biggest difference between DSOs and analog scopes is that the DSO digitizes the input signal and stores it in memory. This means that the input data can be manipulated in a variety of ways. The analog scope shows you a real-time display of the waveform. Here are a few of the major ways that DSOs differ from analog scopes:

- Digital scope displays are almost all color LCD displays rather than CRTs. Further, many of the scope functions on a DSO are controlled through onscreen menus.

- The trigger point on the DSO display is in the center of the grid rather than on the left side. Also, the DSO gives you a wider range of trigger options.
- Because the input data is stored in memory, the DSO has a unique capability that the analog scope does not have. The DSO allows you to view events that occurred before the scope was triggered. This is referred to as *pretrigger capture*. Suppose that a circuit has an intermittent fault. You could use this fault to trigger the scope and then view the condition of different points in the circuit just prior to the fault occurring. This is a very powerful troubleshooting tool for random-event faults.
- Another strength of the DSO is that it can be connected to a computer, or some other test instrument, for further analysis of the data that it has stored in memory.

OSCILLOSCOPE PROBES

The oscilloscope probe is used to couple the scope to the circuit under test. The better the probe, the less loading effect on the circuit. The input impedance of a scope is usually 1 MΩ of resistance in parallel with a low value of capacitance. In order to match the probe to the input of the scope, the probe should be *calibrated* so that its ratio of resistance and capacitance is equal to that of the scope. Most scopes have a calibration output (usually a 1 kHz square wave at 1 V) that is used to set the probe compensation. You should always check the probe compensation at the beginning of each lab session. The capacitance of the scope probe is adjusted so that the scope displays a clean square wave. If the probe is over compensated you will see a spike on the leading edge of the square wave. If it is under compensated, the leading edge will be rounded.

The maximum frequency that you can measure with an oscilloscope is determined by the type of probe that you use, as well as the scope itself. There are several types of scope probes available.

Active and Passive Probes

Active probes contain specially developed integrated circuits in the probe itself. They provide the highest level of performance, but they are very expensive. Active probes can offer a bandwidth as high as 2 GHz and have an input capacitance of less than 1 pF. Passive probes are usually limited to frequencies of less than 500 MHz.

Passive probes are much more common, and in most cases offer a level of performance that is quite sufficient for most applications. The cheapest passive probes are 1× probes. This means that they have no significant input resistance and do not attenuate the input signal. Because of their low input impedance, and rather high input capacitance, 1×probes can load down a circuit under test and limit the bandwidth capability of the scope. 10×probes are better than 1× probes. They attenuate the input signal by a factor of 10 and have lower input capacitance. Many probes can be switched between 1× and 10×. There are also 100× passive probes available for very high frequency measurements, but they are not common and are more expensive. For the exercises in this lab manual, a 10× probe should be used. A comparison of these three types of passive probes is shown below.

	1× Probe	10× Probe	100× Probe
Total input impedance	1 MΩ	10 MΩ	100 MΩ
Input Capacitance	100 pF	10-15 pF	Less than 10 pF

TIME-DIFFERENCE MEASUREMENTS

One of the strengths of an oscilloscope is that it is capable of *comparing* waveforms. Waveforms can be compared by magnitude (voltage) or by time. Figure C.3 shows two sine waves being displayed simultaneously on an oscilloscope screen. Both waveforms have the same amplitude, but there is a *time* difference between them.

The time difference between any two waveforms can be determined using the following procedure:

- Make sure that both waveforms are set to the same voltage reference by setting the input mode selector, in the vertical deflection control section, to the **GND** position. This allows you to set the 0 V reference for each channel to the same position on the grid.
- Set the input mode selector to ac coupling.
- Adjust the VOLTS/DIV controls so that both waveforms are approximately the same size on the display grid. It does not matter if the waveforms have the same amplitude, but it is easiest to make this measurement if the waveforms are close in size on the display.
- Refer back to the two waveforms shown in Figure C.3. As you can see, waveform A crosses the *x*-axis approximately 1.5 major divisions before waveform B. The setting of the horizontal deflection control allows us to determine exactly how much time 1.5 divisions on the scope grid represents. For example, if the horizontal deflection control is set to 0.5 ms/div, then the time difference between the waveforms is found as

$$0.5 \text{ ms/div} \times 1.5 \text{ div} = 0.75 \text{ ms}$$

Thus, waveform A leads waveform B by 0.75 ms, or 750 μs. This is the time difference between the waveforms.

Some final notes before we move on:

- In this example, you compared two sine waves, but the time difference between two waveforms can be determined for rectangular waves, triangle waves, and many other waveforms.

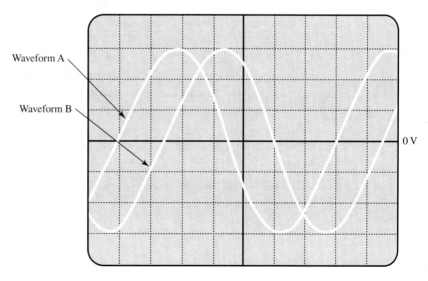

FIGURE C.3   Two sine waves that are out of phase.

- The time difference between two repeating waveforms is only meaningful if the two waveforms have the same frequency. If they do not have the same frequency, then the time difference will change with each cycle, and any time-difference measurement will be meaningless.
- Single-event measurements are another matter. Suppose that a circuit is expected to change its output from 0 V to 5 V some time after a positive pulse is applied to its input. You could monitor the input of the circuit with one channel and the output with the other. The delay time (often called *propagation delay*) between the input pulse and the output transition could be measured in this manner. The question of frequency does not arise.

PHASE MEASUREMENTS

Unlike time measurements, phase measurements normally apply only to sine waves. You have been shown how to determine the time difference between two waveforms. Because of the relationship between time and the degrees of rotation of a sine wave, you can also determine the phase difference, or *phase angle,* between two waveforms.

As you know, one complete cycle of a sine wave rotates through 360°, regardless of frequency or amplitude. For example, each waveform shown in Figure C.3 takes seven major divisions on the horizontal axis to complete one cycle. The phase angle between the waveforms can be found using the ratio of the time difference (in divisions) to the period (in divisions) as follows:

$$\theta = 360° \frac{t}{T} \tag{C.1}$$

where t and T are expressed in *major divisions*. Since waveform A leads waveform B by 1.5 major divisions, you can determine that it leads waveform B by

$$\theta = 360° \frac{t}{T} = 360° \frac{1.5 \text{ div}}{7 \text{ div}} = 77.14°$$

Note that you did not even have to refer to the time base setting of the scope. It is a good idea, however, to set the time base so that the display shows approximately one complete waveform. The more horizontal divisions that a waveform occupies, the better the resolution of your measurement.

To summarize, the steps for determining the phase angle between two sine waves are:

- Determine the number of major divisions of the horizontal axis that it takes for one complete cycle of the waveforms.
- Determine the number of divisions by which one waveform leads (or lags) the other waveform.
- Use the measured times in Equation C.1 to determine the phase angle.

As with time-difference measurements, phase measurements are only valid if the waveforms have the same frequency.

In many instances, a change in circuit frequency results in a change in circuit output. One common reference value that you will often wish to know is the frequency at which the circuit output power drops to one-half its maximum value. This frequency is referred to by several names: cutoff frequency, half-power frequency, and most commonly, -3 dB frequency. Regardless of its name, this frequency can be determined with the help of an oscilloscope. Because oscilloscopes only make voltage and time measurements, however, a little work is needed to determine a frequency that is related to power.

We will use the *RC* low-pass filter shown in Figure C.4a as an example. The *cutoff frequency* (f_C) of this filter is assumed to be the frequency at which the filter's load power drops to one-half its maximum value. As you have learned, load power drops to one-half its maximum value when load voltage drops to 70.7% of its maximum value.

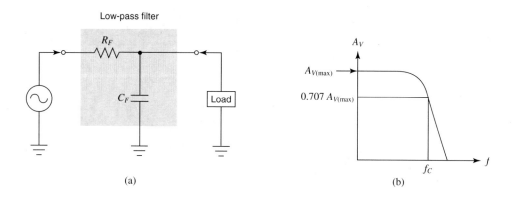

FIGURE C.4

The cutoff frequency of this circuit can be measured using an oscilloscope as follows:

- Set the input frequency to some value well below the calculated cutoff frequency of the filter. This ensures that the output voltage is at its maximum value. Use one channel of the scope to monitor the filter input voltage and one to monitor the load voltage.
- Adjust the VOLTS/DIV control of the channel monitoring the output so that the signal occupies just more than seven major divisions of the vertical axis.
- Now, adjust the variable VOLTS/DIV control (counter-clockwise) until the load waveform occupies *exactly* seven major divisions of the vertical axis (from peak-to-peak).
- Next, slowly increase the input frequency, making sure that the input voltage remains constant. As you approach the cutoff frequency of the filter, the load voltage will begin to drop. When the load voltage waveform drops to five major divisions (from peak-to-peak), you have reached the cutoff frequency of the filter.
- The waveform's period time can now be calculated by multiplying the number of major divisions needed to complete one cycle of the waveform by the TIME/DIV setting. (*Note:* Even though the voltage is no longer calibrated, the TIME/DIV setting is.) The cutoff frequency is the inverse of this value.

At this point, you must be wondering what seven divisions and five divisions have to do with determining the half-power or cutoff frequency for the filter. The answer is fairly simple. If a voltage waveform decreases from seven to five divisions, it changes by a factor of $5/7 = 0.714$, which is very close to 0.707. Thus, when load voltage changes from seven to five divisions on the oscilloscope grid, you can assume that load power has dropped to approximately one-half its maximum value.

Appendix D

Fault Analysis Charts

The **fault analysis chart** is to be used to identify and record fault symptoms and their accompanying analysis. Make as many copies as necessary to complete the fault simulations used throughout this manual. Be sure to record the *fault number* and the *fault description* when using a chart to record the symptoms and analysis of a fault.

Learning to properly analyze a faulty circuit is a skill like any other. As you become more experienced, you will learn how to perform a fault analysis more accurately and quickly. Sometimes, a single measurement is all that is needed to determine the fault and its cause. In other cases, multiple measurements are required. Some measurements are made with a multimeter, while others must be made with an oscilloscope. *You* must learn to determine if you have enough information to properly analyze the faulted circuit and what measurements are necessary.

Let's use the circuit shown in Figure D.1 as an example and fill in one of the charts. The unfaulted circuit has the following values:

$$v_{in} = 20 \text{ mV}_{PP} \quad v_{out} = 3.9 \text{ V}_{PP} \quad A_v = 195 \quad V_B = 3.2 \text{ V} \quad V_E = 2.45 \text{ V} \quad I_C \cong 7.4 \text{ mA}$$

Now, assume that the introduced fault is an **open R_2**. A variety of measurements are taken and the chart is completed as shown on the following page.

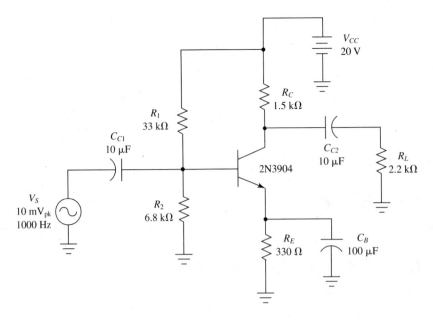

FIGURE D.1 A common-emitter amplifier.

TABLE D.1 **Fault #D.1**

Description: Bias resistor R_2 open
Fault Symptoms: The input signal is unaffected, but the output shows only a slight amount of conduction on the *positive* alternation. There is no output on the negative alternation. V_B increased from 3.2 V to 4.5 V, and V_E increased from 2.45 V to 3.7 V. This means that I_C has increased from approximately 7.4 mA to 11 mA.
Analysis: With R_2 open, the circuit is acting like a base-bias circuit, but with I_C at 11 mA, the transistor is in saturation. With an input signal of only 20 mV$_{pp}$ (10 mV$_{peak}$) it is only capable of bringing the transistor out of saturation for a small portion of the *negative* alternation of the input signal. Since the amplifier is inverting, this explains why there is a slight *positive* signal at the output.

Note: The chart above is intended only as a guideline.

EXERCISE _____   *DATE* _____

Fault # _____   **Description** _____

Fault Symptoms:

Analysis:

EXERCISE _____   *DATE* _____

Fault # _____   **Description** _____

Fault Symptoms:

Analysis:

Fault # _____   **Description** _____

Fault Symptoms:

Analysis:

EXERCISE _____   *DATE* _____

Fault # _____   **Description** _____

Fault Symptoms:

Analysis:

Fault # _____   **Description** _____

Fault Symptoms:

Analysis:

EXERCISE _____   *DATE* _____

Fault # _____   **Description** _____

Fault Symptoms:
Analysis: